HARVEST

THE BODLEY HEAD
LONDON

HARVEST

The Hidden Histories
of Seven Natural Objects

EDWARD POSNETT

THE BODLEY HEAD
LONDON

1 3 5 7 9 10 8 6 4 2

The Bodley Head, an imprint of Vintage,
20 Vauxhall Bridge Road,
London SW1V 2SA

The Bodley Head is part of the Penguin Random House group of companies
whose addresses can be found at global.penguinrandomhouse.com.

Penguin
Random House
UK

First published in the United States as *Strange Harvests* by Viking in 2019
First published in the United Kingdom by The Bodley Head in 2019

www.penguin.co.uk/vintage

A CIP catalogue record for this book is available from the British Library

ISBN 9781847923875

Photograph credits: Page [8]: Lydia Lloyd-Rose
Page [111]: Project Sea Silk
Pages [38, 89, 157, 199, 225, 253]: Edward Posnett

Designed by Gretchen Achilles

Printed and bound in Great Britain by Clays Ltd, Elcograf S.p.A.

Penguin Random House is committed to a sustainable future for
our business, our readers and our planet. This book is made
from Forest Stewardship Council® certified paper.

For Gabriella

CONTENTS

HARVEST

INTRODUCTION

A few years before writing these words, I was working in Canary Wharf in a glass building filled with experts on numbers and systems. As part of my job I wrote reports on companies, markets and political conditions in distant countries. The subjects of my research were serious: million-dollar oil deals in Africa, rogue trading scandals, bribery allegations in the North American construction sector. But looking out of the office windows – or walls – towards north London, I felt none of that gravity, only a dulling sense of the distance between the words I wrote and their meaning.

One of London's two financial centres, Canary Wharf sits on the site of the old West India docklands in a loop of the River Thames. Ships once crammed into its docks to offload rum, sugar and coffee from the West Indies into brick warehouses, now restaurants and flats. The docks thrived for almost two hundred years, until the late 1960s, when the arrival of the container prompted the shipping industry to move farther east. In 1983 most of the land was sold to a developer; within a few years the docks' history was buried under layers of glass and steel, the architecture of a much more liquid capital. Anyone who has visited the Canary Wharf area in east London will know how easy it is to get lost there. To the untrained eye the district fades into itself; the glass buildings on any two streets seem the same, but arranged in a different order.

In 'The Library of Babel', Jorge Luis Borges imagines a vast library containing all the possible 410-page books made of twenty-five characters. In countless hexagonal rooms, four walls of bookshelves hold Borges's imaginary books made up of different permutations of the same letters. Something about my routine in Canary Wharf reminded me of this short story. Each report that I wrote merged with its predecessors; the words, after all, were the same: *shareholder, oil block, concession, acquisition, investigation*; only their order, their syntax, was different.

During my lunch breaks I would go running in the Isle of Dogs, thinking of the ships that once came here and the materials they brought from distant parts of the world: ivory, molasses, wool and hides. The docklands would have smelled different, the odour of these commodities blown by the wind from the Thames. Breathing the air, one might have got a sense of the world growing smaller, each object testimony to new trade links, the speed and reach of the ships of the East India Company. It must have been intoxicating to witness this influx of the exotic, but also disconcerting, as the safe, known world mingled with the unfamiliar and strange.

In London, Antwerp, Venice and other centres of international trade, thinkers tried to make sense of this new world, these smells, sights and textures. Some wrote about these new connections, travelling to distant lands, but others found ways to represent them through objects. Collectors like Hans Sloane, John Tradescant and Ole Worm gathered objects from around the world and placed them in single rooms, or wunderkammers, creating a 'World of Wonders in one closet shut'. More than eccentric collections, these rooms sought to represent the expanding limits of knowledge, to visualise the connections between things and understand networks of relationships. As historian Mark Meadow writes, 'In the world of the curiosity cabinet, as in any other place or period, the strands of blood, trade, authority, and agency were all present, and were all interdependent. To touch any of the strands is to set all the others in sympathetic vibration.'

My favourite wunderkammer belonged to the Danish physician and classics professor Ole Worm. In 1605, Worm set off on a Grand Tour that would last eight years, travelling to Germany, Italy, France, England and the Low Countries. As he went, he gathered objects and met with men of learning, such as the famous collector Ferrante Imperato in Naples. These items ranged from seeds, shells and horn to mirabilia – or marvels – including a 'magic egg' said to have been laid by a woman in Norway. Worm conserved these things, he wrote, 'with the goal of, along with a short presentation of various things' history, also being able to present my audience with the things themselves to touch with their own hands and to see with their own eyes, so that they may themselves judge how that which is said fits with the things, and can acquire a more intimate knowledge of them all.'

During the sixteenth and seventeenth centuries many individuals were collecting objects from the natural world: Athanasius Kircher and Ulisse Aldrovandi as well as Imperato. Children of the Renaissance, they looked largely to classical texts to explain the objects in their collections, tracing their appearance in works such as Pliny's *Natural History* or Aristotle's *History of Animals*. Worm was influenced by these collectors, but he was fascinated by touch, the sensual experience of interacting with objects. In his preface to the catalogue of his collection, the *Musei Wormiani historia*, he wrote of 'the clear intention to lead toward knowledge by direct observation and away from hollow verbiage.' Holding an object, passing one's finger over it, smelling it, one could perceive truth and debunk myths.

I didn't know about the Museum Wormianum while I worked in

Canary Wharf, but had I known, I might have spent hours poring over the catalogue during breaks. Writing in the seventeenth century, Worm found himself at the beginning of a process, the commodification of the natural world, that has reached its apex in our own age. As he wrote, collected, travelled and touched, ships owned by private enterprises, such as the Virginia Company, voyaged to the Indian subcontinent or the New World, trading spices, seeds, tobacco and fibre. Many of the objects in his museum possessed a narrative simplicity, casting a straight line from the 'natural world' to the room of his museum, but others spoke of these complex new trade routes, weaving patterns, contacts and relationships, sprouting branches in their passage from the wild to his own hands.

It intrigued me to think of what Worm would make of our era, a time when almost everything has become a commodity, a tradable object; a time when the exotic is now the domestic, arriving in shops and warehouses as if spontaneously generated. My own work in my Canary Wharf office traced objects' movements, detailing the exchange of materials and things across borders, from the spare parts of aircraft to underwear to oil derivatives. When Worm perceived his new objects, he experienced a sense of rupture with the old ways of seeing, a radical break prompted by encounters with texture. And yet I didn't sense any rupture in my daily routine, only a continual procession of goods, one measured by abstract figures.

It's difficult to measure time when you work in Canary Wharf. The market lives eternally in the present; each unprecedented, unforgettable event swallows the last before it, too, is swallowed and replaced. Then there's the lack of natural light in Canary Wharf. You can remain underground or inside from the moment your train or tube arrives in the station until you finish your workday. I started measuring the passage of time by peering down at the Crossrail station development on the north dock, some fifty metres below my desk, a stretch of water that once held up to two hundred ships from the West Indies. The development began as an enormous cube of empty space below the water level, propped open by thick metal tubes. Gradually the gap filled with cranes, men, blocks of concrete and girders. By the time the cranes reached my line of sight, I decided that it was time to move on.

After I left finance I tried everything and nothing: teaching, archival research, advocacy for a charity, running activities in a homeless shelter. Depending on the day, I felt either a deep sense of relief or a crippling sense of failure. I began reading histories and articles about commodities and extraction, delving deeper into the stories I had skimmed in Canary Wharf. Instead of finding closure, I was left with a vague sensation that by simply existing, eating food, breathing and excreting, I was caught up in stories I did not understand. In his book *Invisible Cities*, Italo Calvino imagines a place called Ersilia where 'to establish the relationships that sustain the city's life, the inhabitants stretch strings from the corners of the houses, white or black or gray or black-and-white according to whether they mark a relationship of blood, of trade, authority, agency.' So abundant did these threads seem in my own life that they merged to form a single dark fabric, where everything was at once connected and separate.

One weekend I visited the London Wetland Centre in Barnes, an artificial wetland hidden within another loop of the Thames on the site of four disused Victorian reservoirs. I found it invigorating to wander along the wetland's paths, stopping to view its lagoons and islets. I saw Canada geese, plovers, mallards and eiders, wildlife from several continents all within 105 acres. I stopped in front of the eiders, large sea ducks that calmly sat on the grass like domestic fowl. Exceptional divers, eiders are most at home out at sea, swimming down to the seafloor for mussels or crabs. It was strange to find them here, wings clipped, marooned in the middle of London, surrounded by murky pond water rather than the smoky brine of the North Atlantic. On a sign in front of the eiders, I read about how for centuries Icelanders have offered the ducks protection in exchange for their lightweight down, a valuable commodity traded all over the world.

I n the 1880s my great-grandfather Robert Posnett took a job in Runcorn, Cheshire, as the foreman of a small tannery owned by his uncle. Until that point, most of my ancestors had been Wesleyan preachers, but Robert found his place among the tanning pits of Cheshire instead of in the pulpit. Practical and hardheaded, he proved well suited to the leather business, the soaking and scraping and drying of flesh, the conversion of bristled animal skin into boot soles, machine belts and animal harnesses. His entry into the trade could not have been timed better. During the Industrial Revolution, demand for leather exploded and Runcorn's tanneries were there to meet it, churning out miles of hard belts for the cotton-spinning machines of Lancashire. Robert later took over the tannery and acquired another in Runcorn, both of which supplied hard sole leather to Allied troops in World War One. By the war's end, he was tanning nine thousand hides a week, converting my own surname, once associated with sermons and pews, into a byword for leather production.

Long before the dominance of synthetics, leather's durability made it indispensable for cotton production and the Allied war effort. Robert's tanneries employed hundreds of workers, providing an important sense of community and belonging in Runcorn. But I could not shake a feeling of discomfort at those figures: thousands of hides a week, hundreds of thousands of hides per year, millions over decades, an immense chain of skin and flesh stretching from the pampas of Argentina to the tanning pits of Cheshire.

The story of eiderdown seemed so different. In its unwritten rules of cooperation, eiderdown harvesting seemed close to the symbiosis that is often said to exist in the natural world between different species: the cleaner wrasse and the reef fish; the plover that picks the crocodile's teeth; the pea crab that guards the blind pen shell. More than a particular harvesting technique, eiderdown offered the possibility of a different relationship with the natural world, one predicated on cooperation rather than domination. The Icelanders and the ducks were equal partners: if either party failed to abide by their contract, the other party could simply walk or waddle away.

Inspired by its promise, I began reading about other harvests that offered similar stories of balance. I gathered seven strange objects: the down of a sea duck, a miniature gelatinous nest, an earthy coffee bean, a strand of dark golden silk, a fine cinnamon-hued fibre, an ivory pebble and an ammoniacal powder. I found it electrifying to hold these strange objects, trying to conjure up the animals or plants from which they came. Their smells and textures were curious and unfamiliar, yet their strangeness transcended their forms, telling stories of cooperation and possibility. Each object served an important purpose in the natural world – as insulator, shelter, waste, anchor or food – yet its removal, its harvest, need not spell discomfort, mutilation or death.

Unlike Worm, I might not have needed to travel far to find my seven: caught up in the fabric of our own lives, many could perhaps be found in supermarkets within a small distance of my home in London or, when I later moved, in Philadelphia. Ferried as they

were by aeroplane or container ship, their presence in cities spoke of
the great reach of the market, its ability to penetrate every nook and
cranny, from the most remote cave to the thickest forest. Traded,
exchanged or gifted, they had passed through countless pairs of
human hands, those of harvesters, traders and shop assistants, before touching my own. What did their presence here mean? What
did it say about our relationship with nature and one another? Handling or smelling them, I sensed that they held secrets, emotions
and promises within their protein barbs, filaments, fibres, strands or
powdered forms. I wanted to pull them apart, in order to reveal their
hidden histories, the lives they touched.

EIDERDOWN

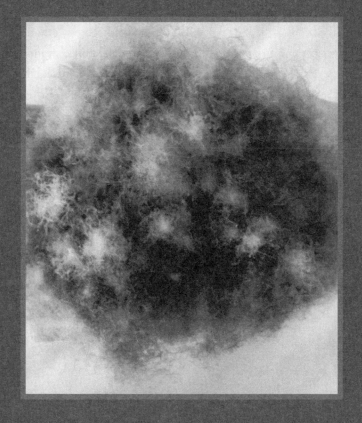

n Ísafjörður, the capital of Iceland's remote Westfjords region, a Lutheran pastor compared eiderdown to cocaine. 'I sometimes think that we are like the coca farmers in Colombia,' he said. 'We [the down harvesters] get a fraction of the price when the product hits the streets of Tokyo. This is the finest down in the world and we are exporting it in black garbage bags.'

It is difficult to describe the weight of eiderdown in a language in which the epitome of lightness is a feather. Unlike a feather's ordered barbs arranged around a solid shaft, under a microscope eiderdown offers a portrait of chaos: hundreds of soft threads branch out from a single point, twisting around one another. Upon each thread are countless small hooks, which allow the down to cling to itself, trapping pockets of air and warmth.

When I returned from Iceland, I asked my wife to close her eyes and put her hands out. After placing a duck-size clump of down in her hands, I asked her what she felt. 'Heat,' she said. She opened her eyes to find the down, a ghostly grey form, hovering above her palms, and pulled it apart, releasing the hooks. It crackled as if electrified, emitting a mild smell that reminded me of burned hair. She scrunched her hands, and the down disappeared in her fingers, compressed into a ball smaller than a duckling's bill.

Over centuries eiderdown has been treasured by those who shared their lands with the eider. The Vikings apparently filled their bedding with eiderdown, medieval tax collectors accepted it as a means of payment, and it's possible that the Romanov tsars, whose dominions were home to some of the largest eider populations in the world, had a taste for the material. Today, its buyers are the global super-rich. In Iceland I heard stories about Gulf royals who sleep under eiderdown in the desert, and Russian politicians whose hearts can be warmed with the gift of an eiderdown duvet.

The properties of eiderdown – extreme lightness and insulation – make sense when you consider the life of its owner. The eider is a fat seabird, more penguin than duck, and many of them spend most of their lives in the Arctic Circle. Visit the Icelandic coast and you will see hundreds, bobbing gregariously in the sea. Brash creatures, their boldness inspires admiration in the Icelanders. 'The eider is an unsung hero, far braver than any bird of prey, which it is known to attack to protect its offspring,' one local told me.

The story of how most feathers come to fill our bedding is anything but comforting. According to trade bodies, most feathers are a by-product of the meat industry. Less fortunate birds may have their plumage ripped out while they are still alive, a practice known as live-plucking (and reported to be common in China and Hungary, both major exporters of down).

As I held a clump of eiderdown in my hand, one Icelander told me something that seems to offer an alternative to this unsettling relationship between humans and birds. He said that the eider it once belonged to is probably still alive. Not in some dark barn, or even in an open-air enclosure, but in the wilds of the Arctic Circle.

In a café in Ísafjörður, the pastor explained how he harvests eiderdown. As part of his parish duties, he runs a small farm, a throwback to earlier times when pastors in remote areas would survive off the land. Even now, life here can be unpredictable, especially in winter, when the weather turns violent. In 1995 two towns in the Westfjords were buried by avalanches, killing thirty-four people. Parish ministers were among the first to reach one of the villages and offer comfort to the survivors.

Every June, he said, about five hundred ducks arrive from the sea and waddle to his farm. Eiders do not naturally nest in such large colonies, but will congregate close to human settlements to seek shelter and protection. The ducks nest anywhere: in tyres, doorways and even houses. 'I always take a lot of flags with me and I put a flag beside each nest so I will be able to find it again. Because they are incredibly camouflaged, these ducks. You can almost step on them.'

At night the pastor guards the flock of eiders from their predators: seagulls, foxes and mink. 'I was quite lucky in that I got interested in guns when I was just a little over twenty,' he said. 'It was before I started studying theology.' If he were to fall asleep, a fox would have a feast of sitting ducks. 'It's more than a financial loss, it's also like they are depending on me. So I don't want to let them down. I used to be a night watchman so I have a little bit of experience staying awake.'

In the Middle Ages, pelicans were thought to pierce their own breast to draw blood to feed their young. The mythical act was known as vulning, a Christlike act of self-sacrifice.

On the pastor's land the eider, too, makes herself vulnerable for her offspring, though it is down, not blood, that she draws from her breast. From this down she builds a nest for her eggs; her own bare skin, freshly revealed, covers them with warmth. She sits on her eggs for some twenty-eight days, during which she may lose a third of her body weight; some mothers may starve to death.

After incubation, the eggs hatch, the mothers waddle back to the sea with their offspring, and the pastor gathers their down, his protection fee. 'I never collect the down until they are gone,' he said. 'Some of the farmers say they like to take a little bit of the down [while the ducks are still nesting]. I just like to leave them, not to disturb them in any way. . . . If you frighten them, they jump up and shit all over the nest.' The 'shit' he describes is not in fact excrement, but a brown oily liquid with an odour similar to that of frying liver. '[It is] so strong,' recorded one Belgian eider enthusiast, 'that an egg touched with it is refused and even discarded by the hungriest dog.'

The scene described by the pastor has been a common sight in Iceland for centuries. Down has been collected here probably since the arrival of Norse settlers in the ninth century. The sight of thousands of tame eiders close to human settlements astounded early European travellers in Iceland. C. W. Shepherd, an Englishman who visited the island of Vigur in the Westfjords in 1862, described a farm besieged by eiders: 'The earthen walls that surrounded it, and the window embrasures, were occupied by ducks. On the ground, the house was fringed with ducks. On the turf slopes of the roof, we could see ducks; and a duck sat on the door-scraper. . . . A windmill was infested; and so were all the outhouses, mounds, rocks and crevices. The ducks were everywhere.'

Environmentalists, economists, and ornithologists have all fallen in love with Icelandic harvesting. There is an irresistible simplicity to the relationship between the harvesters and the eiders. If a harvester cares for the ducks, more and more will come to nest, increasing the amount of down that can be gathered. At times, the relationship can be tested. As the pastor told me, some harvesters cannot resist collecting the down while it is still fresh, removing a portion of an eider's nest before her eggs have hatched and replacing it with straw. It is not pleasant to watch nesting eiders being disturbed, but they quickly return to their nests and apparently do not hold grudges; the same ducks return year after year.

In 1914, Charles Wendell Townsend, a physician and amateur ornithologist from Massachusetts, became obsessed by the Icelandic method while investigating the declining numbers of eiders in Labrador, Newfoundland and Nova Scotia. Dismayed by the practice of duck hunting in North America, he proposed the introduction of eider reservations where down could be harvested for profit. He had read about Icelandic harvesting in travellers' accounts and dreamed of introducing a utopia of ecology and commerce along the coast of the United States and Canada. 'The cooing notes, so long few or absent in many places, would again resound over the waters,' Townsend wrote in 1914, in 'A Plea for the Conservation of the Eider'. 'And best of all, to the practical minded, the birds would pay well for their protection by gifts of eggs and of valuable eiderdown.'

I wondered how fragile the relationship is between Icelanders and the ducks, and whether the drive for profit might somehow kill the golden goose. If demand for down reaches the level of that for rhino horn, bear gallbladders or elephant ivory, could the eider be driven to extinction? Might Icelanders have to develop intensive farming techniques to increase the supply of down, possibly to reduce illegal hunting?

These possibilities seem remote in Iceland, where eiders have been respected for centuries. I saw this first hand on the remote island of Vigur, which looks out into the Arctic Circle. I met a sixteen-year-old boy whose family has inhabited a farm on the island since the nineteenth century. On his forearms I noticed dozens of small scars, from wounds inflicted by puffins, which he catches, mid-flight,

with a vast butterfly net. He was happy to snap hundreds of puffins'
necks to kill them for their meat, but the eider was untouchable, a
sacred bird. His uncle later explained that eiderdown accounts for
more than a third of the family's income.

If the Icelanders' relationship with the ducks does change, then
it may be for different reasons. The Norwegians once harvested
down all along their coastline, but in the 1960s they discovered a far
more profitable natural resource – oil – and began to move away
from remote coastal areas. 'The eiders tried to move with them because they felt protected by humans,' explained one harvester. 'The
ducks would rather live with cats and dogs than seagulls.'

Iceland has not yet struck oil, but it does have untapped reserves
of hydroelectric and geothermal power, which the government is
looking to exploit and export. (It is considering plans to lay a submarine cable between Iceland and Scotland that would provide
electricity for European homes.)

The future of Iceland's renewable energy sources rouses strong
feelings in the country, where large swathes of pristine landscape
have already been flooded, or 'drowned,' to create the conditions
required for hydroelectric power generation. If the Icelanders choose
to develop their country's renewable energy sources to the full and
export electricity as the Norwegians export oil, eiderdown harvesting may begin to seem an outdated tradition.

In the cycle of a natural resource, it is generally its extraction
that is its most unpleasant and destructive stage. To win rights to an
oil block in a developing country, an oil company may slip a sweet-
ener into an official's pocket. To increase the production of feathers,
a goose farmer may find it profitable to wrench them from a bird
while it is still alive. The story of down harvesting is so different

from these narratives of compromise and degradation. Once down is traded and marketed, however, it becomes just like any other commodity.

'What looks like a nice, peaceful, old-fashioned small trade from the outside is actually an octopus of monopoly and manipulations,' said Jón Sveinsson, an Icelandic businessman. 'Scratch the surface, follow the money, and the picture quickly changes from idyllic hobby to cut-throat exploitation.'

A former naval officer, Sveinsson has devoted his life to eiderdown. Unlike many harvesters, he is involved in all stages of its life cycle, from the duck's nest to an oligarch's bedding. Since childhood, he has harvested down on his family's farm in the Westfjords, later investing hundreds of thousands of euros developing his own machinery to dry and clean it. Above all, he is a marketing man. An avid reader of Edgar Allan Poe, he compared his marketing efforts to detective work, 'looking for traces of potential clients, finding what spots they frequent, like a hunter tracks a rare game to its watering hole.'

Sveinsson explained that to produce a single kilo of refined down takes a colossal amount of work. About sixty nests must be collected, dried and cleaned to remove dirt, seaweed and vegetation. This process was mechanised in the 1950s when Icelanders developed their own drying and cleaning technology. Despite all this labour, the value of the down produced and refined in Iceland is only a fraction of its retail price. Intermediaries, such as Japanese and European wholesalers, buy in bulk after it has been processed. It is then stuffed into pillows, duvets and clothing sold in Japan, China, Germany and Russia. The Icelanders receive about 3 million euros for the three tonnes of down that – give or take – they export each year, but according to Sveinsson, its retail value may be ten times this amount. 'The harvesters actually get a lower percentage of the retail value than an African coffee grower,' he says.

His words reminded me of what the writer Rebecca Solnit called (in a 2008 essay for *Harper's Magazine*) Iceland's 'fairy tale told backward', in which its people 'had been dispossessed of their great gifts and birthrights'. It began, she argues, about three decades ago, when Iceland privatised the right to fish and introduced quotas that could be traded and accumulated. Today large firms predominantly control the fishing industry. Then in 2006 a remote highland wilderness in eastern Iceland was flooded to create a huge reservoir as a power source for an aluminum smelter. The cost of the hydropower project was about $2 billion, much of it borrowed from international banks. Critics of the plan claim that the environmental and financial price was too high for the several hundred jobs it created.

Eiderdown, sold off cheaply, seemed to be just another chapter in this story. The writer Andri Snær Magnason points out that the Icelandic word for 'windfall' is *hvalreki*, literally a beached whale. While a windfall in English evokes falling fruit, its Icelandic translation is a stranded sea mammal, a gift of free meat.

When explaining why the Icelanders are getting such a raw deal from down, Sveinsson pointed to Iceland's historical relationship with its largest export, fish. 'Abundant fishing grounds have been a resource that have shaped the Icelandic economy as well as our mentality, a blessing and a curse in one,' he said. 'A kind of hand-to-mouth attitude developed – there was always the knowledge that one day the weather would calm down and the boat could set off for another good catch.'

*

n myths, fables and hagiographies, one often reads of the ability of individuals to tame wild creatures, burnishing their reputation for virtue, sensitivity or holiness. It is said that Saint Cuthbert, the seventh-century missionary who settled on the Farne Islands off the Northumbrian coast, protected and tamed eiders. (Today in Northumberland the eider is sometimes called Saint Cuthbert's or Cuddy duck.) Many of these stories were built upon embellishment or pure fantasy, but in Iceland, travellers' accounts repeatedly confirmed the existence of this strange relationship between Icelanders and the ducks. Writing in 1875, the English explorer Richard Burton commented that the eider was a 'barn door' bird as 'tame as horse-pond geese'. 'No salute must be fired at Reykjavik,' he wrote, 'for fear of frightening "so-materia mollissima."'

I began reading all I could about eiderdown, scouring books and articles, devouring obscure travellers' accounts and biological papers, contacting eiderdown harvesters. I wanted to know how this relationship worked, how it was possible for a wild bird to behave as if it were domesticated. How did this strange tradition come about? How was it that the relationship had been preserved despite the arrival of the market? Could you simply transfer the Icelandic tradition to another coast? Could eiderdown harvesting teach us about our relationship with other species? If earlier writers had looked to Icelandic harvesting with a sense of sublime wonder, then I sought it with a frenzied gaze, looking for answers.

*

Back in Ísafjörður, the pastor told me about a Japanese film crew who had made a documentary about him. For several weeks they followed him around his farm, recording him collecting down with his children, while sidestepping nesting eiders. He appeared bemused by their attention, just as he was by mine. After all, he said, the down was only some *brauð*, a slice of his daily bread.

The pastor's parish lay a short drive from Ísafjörður in another fjord, Önundarfjörður. It was once reached via a winding mountain road, but since 1996 the two fjords have been connected by a vast tunnel that bores directly through the hillside. One morning I emerged from its confines, blinded by the subarctic light, and headed to the pastor's church to take a tour of his nesting area.

Like many Lutheran parishes, the pastor's land is among the most prized in the Westfjords. In the shadow of a steep glacial wall, his family home and church overlook a floodplain that leads to the shoreline. After ten minutes in the tunnel's darkness, I was overwhelmed by the intensity of the colours in the fjord. In the subarctic light the tones seemed almost hyper-real, the blues of the sky and sea merging to form a single aquamarine. Against the white of the fjord's beach, the pink of a buoy stood out, dabbed as if it were an afterthought.

It was hard to conceive of a more peaceful spot for a family, yet there was an air of abandonment to the pastor's home. Children's toys lay scattered around on the floor, a layer of dust covered the work surfaces, and his books on wartime heroes and mountaineers lay untouched on a shelf. The previous winter, he explained, the weather had got so bad that he and his family had to leave the parish and take an apartment in Ísafjörður. He now rarely spends any time in the house. 'We were trapped here for thirty-six hours with no electricity, no phone,' he said. 'I can't be responsible for that.'

As if to prove a point, he showed me the damage wreaked by the blizzards: a mangled fence, the scoured surface of a barn, and the bent cross on top of the church, more weather vane than symbol of devotion. In the distance I could make out Flateyri, the fishing town at the mouth of the fjord that was partly destroyed by the 1995 avalanche. Behind it loomed a vast avalanche barrier, a constant reminder of the perils of living in these remote fjords.

I was late for the eiderdown season, but the pastor offered to lead me around the land and re-enact his summer ritual as he had done for the Japanese film crew. Dressed in an old Polish military uniform, he headed off across the flat plain towards the beach, hunting for any nests that he had missed. It was a still day, the silence broken only by the call of an oystercatcher, alarmed by our presence.

Walking on this flat land, I felt as if I'd missed out on a great gathering. All around us were hundreds of small piles of crushed mussel shells, the remnants of the eiders' feasting, ground up by their powerful gizzards. Quartz-like, these remains had an understated beauty, glinting in the light. 'In the later part, everything is going crazy,' the pastor said. 'Birds and chicks running around. Arctic terns attacking all the time. It's good to have a broomstick.'

As we walked towards the beach, I commented that, aside from the buoy, there was practically no rubbish in the fjord. 'There is very little,' the pastor replied, 'but sometimes something interesting pops up. Like two or three years ago I found this green cylinder.' He led me over to a discarded wooden reel, the kind used by industrial trawlers, reached underneath it, and pulled out a bright green plastic cucumber. 'You know, I thought, "What is this? Is it a World War Two bomb or something?"' He unscrewed the top and unsheathed a rubber dildo.

Among the excreted mussels and sea plants, a dildo was the last thing I had expected to find washed up in this remote fjord. 'What I'm guessing,' the pastor said, 'is that somewhere there was a big container of sex toys that just washed overboard.' Stranded en route to its fleshy counterpart, the flaccid rubber belonged to a different world from that of eiders. And yet it would outlast all the other objects on the beach; while eiderdown rotted, ordnance rusted and wood decayed, the dildo's rubber would endure, a constant reminder of some maritime mishap and the colossal power of Arctic tides.

Dildo sheathed, we made our way back to the church, counting the pieces of mussel quartz. Suddenly the pastor let out a cry and pointed to a nest that he had missed during the previous gathering. Covered in moss, grass and broken eggs, it looked like a furry grey omelette or pancake. He wedged his stick under the down, easing it gently from the grass, and picked it up. Laden with seaweed, twigs and dirt, it reminded me of the contents of a vacuum cleaner, half fluff, half debris. Unlike the clean down my wife had held, it had a pungent, mouldy aroma, suggestive of the sitting duck from which it came. Looking closely, I saw the remains of several eggs caught up in the down. Rendered rubbery by rainfall, their fragments were proof of what the pastor had said; he always allowed the ducklings to hatch before collecting their bedding. 'Take it as a gift,' he said.

Some fifty years before I visited Iceland, the English writer Gavin Maxwell became momentarily possessed by Icelandic harvesting while living in Sandaig, a bay in remote western Scotland. He began to read about harvesting and even travelled to Iceland on a research trip, contacting the renowned Icelandic ornithologist Finnur Guðmundsson. Maxwell is best known for his relationship with his pet otters, but he also adored eiders, writing that they reminded him more of 'animals than birds'. 'Perhaps,' he mused, 'it is their peculiarly uneven voices, or the way their massive bills ascend in a

straight line to the top of their skulls without any "scoop" in between. Or perhaps it is their curious and very individual smell,
which seems as if it could have nothing to do with a bird.'

Unlike previous travel writers who described eiderdown, he
dreamed of re-creating the success of the Icelanders. From his cottage he'd noticed eiders nesting on a small island close to his home;
he counted only thirty pairs in total, but he imagined that he might
turn it into a duck magnet, drawing thousands of ducks like iron
filings. Inspired by Icelandic methods, he proposed the erection of
flags, wind propellers, scarecrows, artificial nesting sites, wind shelters and decoy drake eiders to attract the ducks. He concocted plans
to feed the ducks with mussels and – apparently – even to entertain
them with music.

Maxwell is most famous for his memoir *Ring of Bright Water*,
which recounts his domestic life with several otters, but during
much of his life his relationship with the wild had been a bloody
affair. After his wartime service, he bought a small Scottish island,
Soay, and set up a basking shark fishery. He hunted these great fish
with a harpoon, colouring the Scottish waters red with blood, and
later writing about his experiences in *Harpoon at a Venture*. Laden
with gory detail, the book is a celebration of one man's war against
these large, placid filter-feeders. 'In it,' writes Robert Macfarlane,
'Maxwell – Queequeg with an Oerlikon, Ahab with an inheritance
– demonstrates an awesome lack of interspecies empathy, and a
broader insensitivity to affect that verges on the psychopathic.'

In the later years of his life, Maxwell saw eiderdown as something different. Like so many writers, he wrote of harvesting in utopian terms, describing how early settlers had lured eiders with 'an
elaboration of fluttering flags, little wind-driven clacking propellers,
and reeded wind instruments that would sigh, groan or trumpet,

according to the strength of the breeze.' The Icelanders, he believed, had some sort of special ability to understand ducks that verged on the mystical. While he took copious notes on Icelandic techniques in an attempt to understand Icelanders' success, he allowed that there was 'a certain *mystique*' about the practice. 'Some men,' he wrote, 'seem to have "green fingers" with Eiders.'

When I visited the pastor, I was infused with Maxwell's own vision of down, a contrast to stories of degradation and exploitation. Patient and sensitive, the pastor seemed to fit perfectly Maxwell's vision, heroically guarding the ducks, never taking the down too early. And yet I had an itch. In the nineteenth century, British, French, Scandinavian and American travellers, many of them scientists, started visiting Iceland, which was then a Danish colony. Wandering through its volcanic landscape, admiring the frames of its miniature horses and its colossal populations of eiders, or listening to the Icelandic tongue, they described Iceland in magical terms. Fed up with their descriptions of 'scenes of thrilling horror, of majestic grandeur, and of heavenly beauty', the explorer Richard Burton wrote of a condition called 'Iceland on the brain'. I feared that I was falling into a similar trap, failing to write about another character in the drama.

The Arctic fox – *Vulpes lagopus* – is the only mammal in Iceland whose presence predates the island's settlement in the ninth century. Much smaller than its red cousin, the Arctic fox has miniature soft paws that are entirely covered in fur (hence its species name, which means 'rabbit foot' in Latin). Perfectly adapted to the country's varying seasons, the foxes can roam thousands of kilometres, their winter coats transforming from white to brown or even blue in the summer months. It is now believed that these nimble animals migrated to Iceland across the Arctic sea ice from Greenland or northern Europe thousands of years ago in the Holocene.

When settlers arrived in the ninth century with domestic animals, it was inevitable that they would clash with the Arctic fox. Fearful of damage to livestock, farmers clubbed, trapped and later poisoned and shot them. Such was the fear of the foxes among Icelanders that a myth emerged that they had been sent by the king of Norway to punish settlers for their abandonment of their motherland. Legislation to encourage their extermination dates back to 1295, and even today each community in Iceland is obliged to hire a fox killer to visit all known dens in the area to exterminate their occupants.

As I drove back from the pastor's farm to Ísafjörður, it struck me as unlikely that the pastor, this kind man who refused to disturb an eider, was a true fox killer. Despite the military uniform, glasses and talk of guns, he never actually described killing a fox. I later asked his friend and neighbour, a long-time eider farmer, who really kills the foxes in the fjord. 'I do,' he said with a smile. 'The priest has never shot a single fox.' I wondered whether I had not experienced a sanitised version of down harvesting, one primed for the Japanese market. Curious, I asked around in the Westfjords, searching for the name of an effective fox killer. Again and again the same name popped up: Valdimar Gíslason. Famed throughout the Westfjords for his cunning and ruthlessness, he had raised a private army, a band of killers to protect their capital of eider ducks. Eider farmers spoke of him in hushed tones, revering him for his dedication to fox extermination.

Valdimar lived in Dýrafjörður, the fjord beyond the pastor. The following week I visited him at his home, a sparsely furnished

farmhouse that overlooked the fjord. His Danish wife, Edda, welcomed me in, the sound of choir music blending with the roar of an
eiderdown dryer in a barn outside. In his seventies, Valdimar introduced himself in broken English, gesturing when words failed him.
'Do you know Icelandic sagas?' he asked me, pointing towards a
bookshelf with a complete edition. 'The most beautiful saga is
Snorra Edda,' he said, smiling at his wife.

A retired maths teacher, Valdimar has lived in the fjord for seven
decades. Since his boyhood, he has gathered eiderdown from his
land, once the largest producer of down in Iceland, and pursued the
foxes that wish to dine on his ducks. 'Sixty-five years, fifty nights
every spring,' he said, as if considering a mathematical problem.
'You can tell me how many nights I have stayed awake.' After so
many nights of shooting, his hearing had begun to fade.

Every spring Valdimar assumed his title of *gran comandante*,
assembling a private army of neighbouring eiderdown farmers. In
accordance with his written orders, they head out in cars, armed
with rifles and walkie-talkies to hunt foxes. A good haul, Valdimar
said, is two tails per week. 'We killed fourteen foxes this spring,' he
told me. 'Last year eighteen.' A sense of competition keeps the team
awake; no one is allowed to turn in until eight or nine in the morning. 'He goes to sleep when I get up,' Edda said, as if her husband
were a shift worker.

Maxwell wrote that the Icelanders' relationship was based on the
ducks' aesthetic sensibilities, their preference for certain objects and
colours. Yet Valdimar's success was predicated on a different logic.
More gamekeeper than duck whisperer, he was able to preserve one
species simply by destroying others. 'One spring [decades ago] we
killed three hundred ravens,' Edda told me cheerfully. 'We used

drugs and they fell asleep. He [Valdimar] came and hit them on the head.'

The practice of poisoning or sedating eiders' predators was once widespread in Iceland, carried out by farmers and government officials alike. Such techniques may have saved the lives of countless ducks, but they have historically proved highly destructive to the island's ecosystem. In the early 1980s, for example, more than four thousand ravens per year were killed under the country's pest control programme. Worse, the disappearance of Iceland's white-tailed eagle, which almost went extinct in the 1960s, has been attributed partly to the activities of eiderdown farmers. Although eagle numbers have now recovered, their torrid history speaks of what Andri Snær Magnason, himself an eiderdown farmer, calls the 'dark side' of the eiderdown trade: however virtuous harvesters may be, they have a strong incentive to kill any species that threatens the prized bird.

After a night of watching and shooting foxes, Valdimar and the other fox killers retire to his house to drink coffee, eat cake and recite poetry. 'Every morning,' he said, 'everyone waiting must come here with a poem. He must read it for us. Poem after each night. About the foxes, flowers, ducks, eiders. Many about the foxes with names.' Valdimar brought out a binder thick with a collection of poems, the product of countless nights' work. Browsing through the collection, I found most of them seemed to be about foxes, the hunters, rather than ducks. 'We don't hate them [the foxes],' he said. 'They are trying to live.'

In general we try to dehumanise those beings we seek to destroy. In the words *vermin* or *pest*, the different hues of a living being are condensed into a single colour. But Valdimar and his companions

deliberately set out to breathe life into the foxes even as they took it.
They gave the foxes names, represented them in writing, joked
about them, and spoke of them as family pets. I looked up and saw a
gravestone hung on the wall with four footprints and a photo of a
dead fox named Kusi. 'We have one hundred poems about this fox,'
Valdimar said fondly. 'It could spring over the hedge. The only fox
we have seen [do that].'

As Valdimar waxed on about the foxes he had known, it was hard
not to feel that he preferred them to the ducks he guarded. Unlike
the eiders, the foxes made heroes of farmers, turning them into pro-
tectors. Stripped of the fox, perhaps eiderdown harvesting might be
a mundane activity, a form of rent collection. But there was another
reason that explained Valdimar's fondness for foxes, one I appreciated only once he brought out his notebook. Bound in red leather,
it recorded the number of ducks that had come to his farm over five
decades. In the early 1960s, it recorded over six thousand ducks, but
gradually the figure began to decline. Today, Valdimar said, there
are only one and a half thousand nests, a fraction of his original
haul. I guessed that the foxes were to blame, but he replied that the
fox was innocent; it was his neighbours, his fellow hunters, who had
reduced his take of eiderdown. 'Once we were the only colony [in
the fjord],' Valdimar told me. 'Now there are five or six. The birds
went from us to the others.'

When settlers first came to Iceland from Norway in the ninth
century, fleeing political turmoil, they were confronted with a harsh
subarctic landscape, devoid of large animals. Unlike other settlement societies, they did not encounter any indigenous communities
whom they could imitate, but largely sought to replicate structures
in Norway, bringing their own domesticated animals and crops.
In addition to these resources, settlers foraged along the island's

extensive coastline for driftwood, molluscs, edible plants, eiderdown and beached whales. Such a windfall offered opportunities, but also potential for great conflict. Indeed, much of early Icelandic law deals specifically with harvesting rights for these scarce resources in order to avoid violence.

The sagas piled on Valdimar's shelf, those bloody tales of Iceland's settlement period, recount how huge disputes would erupt between settlers over driftage rights, particularly over beached whales. In *Grettir's Saga*, for instance, several chieftains and their bondsmen rush towards a beached whale and start a brutal battle to harvest its meat, wielding butchering axes, knives, whale ribs and flesh. 'The fighters kept exchanging lethal whale-meat missiles,' writes the saga's author. 'That's how these boors play the game of battle.' Like the beached whale, the ducks in Valdimar's fjord were a finite and highly valued resource; any gain by a farmer in eiderdown would be offset by another's loss. Yet it was impossible to conceive of any conflict over their down, to conjure a scene in which they ran towards the nests, turning their rifles on each other or making missiles of eider eggs. The fox had masked the competitive logic of the harvest, uniting the farmers in common cause, their quest to avert a duck massacre.

n 1965, Gavin Maxwell embarked on his project to lure eiders to his island. He wrote to ornithologists, the Scottish Council for Development and Industry, the Scottish Wildlife Trust, the Royal Society for the Protection of Birds (RSPB), the conservationist Sir Peter Scott, the Nature Conservancy and a firm of feather traders for advice and to enlist their support for his eider project. The response was mixed. While offering his support, Harry Milne, an expert on eiders at the University of Aberdeen, told Maxwell of his 'serious doubts . . . whether you can simply transfer a "situation" from one place to another and expect it to work without having the intermediate evolutionary steps.' Undeterred by such practical questions, Maxwell pressed on, spurred by something powerful and nameless. 'I am going ahead with the project,' he wrote to the RSPB in 1965, 'and that, so to speak is that.' He secured permission from the National Trust for Scotland to use the island, and set aside a budget of £2,100 of his own money for the experiment.

In autumn 1967, after years of research, Maxwell began to prepare the island to receive thousands of ducks. Two of his assistants set the island on fire to clear it of any thorns and brambles that might impede the ducks' passage; catchments for rainwater were dug so that the island's guests would not suffer from thirst. The final touches were put on the island in March 1968, a few months before the ducks were due to arrive. Maxwell's friend Richard Frere covered it with a hodgepodge of objects to lure the ducks and deter gulls: little green tents, coloured bunting strung between poles, bells and tiny windmills (the ducks, Maxwell told Frere, were known 'to be fascinated by the sounds of bells and whirring windmills'). It was, Frere later wrote, 'a virtual reproduction of an Icelandic colony', but come summer, one thing was missing: the thousands of ducks that Maxwell had envisaged.

In his memoirs, Maxwell doesn't dwell on the absence of the eiders, though he confided to Frere that the ducks may have been put off by the parched state of their lodgings. The previous autumn, Maxwell's assistants had let a fire get out of control while clearing the island of thorns and brambles. According to Frere, it raged for two days, turning the island into a blackened desert. 'We hoped that spring would bring a strong growth of young heather,' he wrote, but 'by March a depressing sterility was still in evidence.' Maxwell explained that the eiders would have felt 'over-conspicuous and insecure'.

Maxwell blamed the fire, but I wondered whether his plan would ever have worked. Talking with Valdimar, I appreciated that our own part of the bargain was to efficiently and swiftly kill any threat to the ducks and to do so over a long period. By contrast, Maxwell viewed Icelandic harvesting as a technology, one that could be bottled, shipped and reproduced, rather than a relationship based on systematic killing. Despite his ambitions, he never got the chance to repeat his experiment; he died of cancer the following year, leaving behind a raft of unfinished projects: a book on wild mammals, the construction of a small zoo, and his cherished plan to draw eiders to the Scottish coast. 'The magic that had once glossed the world of Camusfeàrna [Sandaig] had been wearing thin for a long time,' he wrote shortly before his death. 'Now only the base metal of mistakes showed through, bare and ugly.'

It is easy to read tragedy in Maxwell's inability to re-create Icelandic techniques, but I also felt some relief. Set on feeding the ducks or serenading them with music, he wanted to shift the balance between the eiders and humans, bringing the ducks one step closer to domestication, much like the wild otters with which he shared his home. Yet I also felt a loss when I read about Maxwell's failure. I did not sense any cruelty in any of the harvesters whom I met, merely the simple unsentimental coldness of a people who lived off the land, who killed or starved. All this said, part of me still preferred Maxwell's version, his image of Icelanders with 'green fingers'. Walking through the Hornstrandir reserve, the most remote area of the Westfjords, I would often see Arctic foxes patrolling the banks of rivers or the coastline, looking for birds and eggs. At night

I'd listen to their high-pitched calls from my tent, my dreams populated with images of small children at play. Like the birch trees that once covered Iceland, the foxes were a reminder of what this island was like before the arrival of human settlers in the ninth century. To kill them was to assert a certain logic that felt at odds with the promise of eiderdown, its essential balance, and I began to hunt for places in the Westfjords that might fit Maxwell's image.

Attached to the mainland by a small flap, the Westfjords peninsula almost seems to belong to Greenland, which lies about two hundred miles to the north-west. Geologically, the peninsula is much older than the rest of Iceland, lacking the country's spectacular volcanoes and swirling lava flows. Rather than heat or fire, its deep U-shaped valleys speak of the colossal power of huge glaciers, which gouged its rock to form deep fjords that extend outwards like fingers. Coveted by the Vikings and eiders alike, they offer protection from the harsh weather, but not from the Arctic foxes, which can reach any spot in the peninsula with their rabbit paws. The exceptions are small islands in the Westfjords, such as Vigur and Æðey. 'The islands of Vigr and Œdey . . . are [the eiders'] headquarters in the north-west of Iceland,' wrote C. W. Shepherd. 'In these they live in undisturbed tranquility. They have become almost domesticated and are found in vast multitudes, as their young remain and breed in the place of their birth.'

In writings on eiderdown, the island of Æðey has assumed an almost pre-eminent status among nesting sites because of the vast size of its eider population. In the summer months thousands upon thousands of eiders besiege the island, coating its grassland in brown and white hues, the sound of cooing drakes almost deafening the visitor.

'The Arctic fox kind of pushes them [the ducks] to the island,' Alexíus Jónasson, one of the island's owners, told me when he picked me up from the shore in his small boat. 'They find more peace here.' Such is the size of the harvest that 66°North, the Icelandic clothing manufacturer, has offered jackets filled solely with Æðey's eiderdown.

It took me some five hours by car to reach the pick-up point to Æðey. A mere twenty-four kilometres of water separates it from Ísafjörður, but the coastal roads of the Westfjords wind like furled ribbons, tracing the outlines of the fjords' fingers. Driving along them, you seem to brush your destination, only to fold back on yourself, and repeat the process. By the time I pulled up at the pick-up point, overlooking the island, I was trying to figure out the amount of petrol I had burned in reaching this remote place, my discomfort assuaged little by the hitchhikers I had picked up, a young French special-needs teacher and an American with Icelandic citizenship who believed himself to be the next pope.

In his mid-thirties, Alexíus spoke perfect English, honed from years of reading Tolkien, Clancy and the instruction manuals for fish-processing machinery from his workplace in Ísafjörður. The sixth of seven generations who have lived on Æðey, he gathers down together with his two brothers, Jónas and Magnús. But Alexíus told me that it is their mother, Katrín, who is really in charge of the down harvest, a reminder of the key role that women and family matriarchs have traditionally played in the trade, from nest to bedding. (Indeed, I later learned that a savvy businesswoman named Erla Friðriksdóttir is the largest processor of down in Iceland, trading about a tonne per year, a third of the country's entire production.)

Handling the boat with expert care, Alexíus guided us across the narrow stretch of water, scattering the seabirds on its surface. As we approached the island, I could hear the cries of his children playing in the water under the gaze of his mother. 'It's nice for them to be free,' Alexíus said, before we heard a high-pitched wail. 'We get the occasional argument.' In the eighteenth century the French naturalist Georges-Louis Leclerc remarked on the differences between those who harvested eiderdown in Iceland and those whom it warmed: 'In that rough climate [of Iceland] the hardy hunter, clothed in a bearskin cloak, enjoys in his solitary hut a peaceful, perhaps a profound sleep, while, in polished nations, the man of ambition, stretched upon a bed of eider-down and under a gilded roof, seeks in vain to procure the sweets of repose.' Leclerc's words, I reflected, seemed to stand the test of time. Observing Alexíus and his family, I struggled to envision them wearing Æðey's down jackets, which can cost $8,000 each; Alexíus, for his part, preferred a faded grey boiler suit.

The buildings on Æðey give it the appearance of a livestock farm: there was a barn for sheep and a loading bay from which they could be transported to Ísafjörður to be slaughtered. We headed over to this season's eiderdown, two huge grey squares of bedding laid out in front of a barn to dry. Watching Alexíus and Magnús turn it over with their hands, I felt the urge to lie on it, to nestle in the down. Alexíus's daughter, an eight-year-old with strawberry-blond hair, joined us. 'We have rare bits!' she announced quite unexpectedly. 'Rabbits,' clarified Magnús. 'We have rabbits.'

I followed Alexíus as he wandered across the island, holding up a stick to distract terns that were dive-bombing to protect their chicks. No idle measure, the precaution saved us from the powerful strikes of their beaks, which are sharp enough to draw blood. He whistled, and a group of Icelandic horses, slender and shy, trotted towards us from the other side of the island. 'They are actually quite fond of going over to the nests and eating the eggs,' he said. 'They break the eggs all over the eiderdown.' We headed onwards to the island's lighthouse, studiously avoiding puffin nests in the fissures that riddle the island. At the island's tip we sat down on a rock, admiring the spouts of several humpbacks, there to feed in Iceland's rich waters after journeying from their breeding grounds in the Gulf of Mexico.

When Maxwell wrote about eiderdown, I imagined that he had Æðey in mind as a model. Watching the whales' spouts rise and dissolve in the distance, it was easy to believe this place a rural utopia, a place where eiders could nest in peace and children roamed alongside geese, rabbits, puffins and horses. Around us life exploded from the water, the skies and the crevices in the rocks. All one had to do was to observe it, wait patiently and gather eiderdown. But then we were met by Tása, the family dog, whose job it was to catch any mink that swim over from the mainland. 'She's a gentle family dog,' Alexíus said, 'but when she meets the mink she goes ape shit. It's quite messy when she gets it. She starts one end and breaks every bone.'

The recent history of the Westfjords is really the story of rural depopulation, of a vanishing culture next to the Arctic Circle. Over the past decade, countless farmers have packed up and left the region, tired of the weather, isolation and poor roads. The region's tunnels and bridges, intended to increase mobility, have served as escape routes, emptying the fjords of Icelanders. 'The government is always making it harder for people,' Magnús said. 'There's so little money in it, being a farmer, it's becoming a lifestyle choice.' Unlike many livestock farms in the Westfjords, eider farms are still populated, their down a source of stable income. Faced with rising costs and falling profits, the brothers stopped sheep farming in 1990. Instead of farming, they chose to specialise in eiderdown, shepherding ducks instead of sheep. 'It's like a family thing,' Alexíus said. 'Everybody helps out.'

In *The Botany of Desire*, Michael Pollan's inquiry into our relationship with the apple, potato, marijuana and tulip, he asks the question 'Who is domesticating whom?' Each of these plants, he points out, has evolved to gratify human desires, shaping our own lives: 'All these plants, which I'd always regarded as the objects of my desire, were also, I realized, subjects, acting on me, getting me to do things for them they couldn't do for themselves.' Reading his words, I thought about the farmers' relationship with the eiders. While the Icelanders manage the eiders, drawing them to their dwellings with the promise of protection, the ducks had also bound the farmers to these remote corners. 'It's like the ducks are keeping us here,' Magnús told me. 'We wouldn't be here if it wasn't for them.'

Yet there was another presence, one that silently haunted the harvesters' imaginations, fed by novels or television: the men and women of ambition who wore the jackets or slept under the duvets. Unlike Icelandic wool sweaters, yogurt or finance, eiderdown has for much of its history been an international commodity, first traded by the Danes, then – after Iceland's independence – bought by middlemen in Europe and Japan. Far from isolated, Iceland's three hundred and fifty eider farms are connected to distant metropolises, an invisible thread running between the ducks, the harvesters and the buyers, much of whose wealth, no doubt, is built on a form of resource extraction that is far removed from the gentle rhythms of down gathering.

In the course of this book, I spent time with many remote

communities, cut off from urban centres by the harsh contours of landscape, stretches of water or walls of rock. I often found that their encounter with the market was a source of trauma, leading to a collapse in the old ways of living and seeing, *a great transformation*. But I felt no quake on Æðey or indeed on any of the other eider farms I visited. Despite rumours that the Chinese were buying up eider farms, there was little sign of eider speculation upsetting the pattern of land ownership. Every season, the ducks came, and their down was harvested and shipped off to the middlemen for a stable price, much as has been done for centuries.

Historians and sociologists have long marvelled at the continuity of Icelandic culture since its settlement. According to the sociologist Richard Tomasson, 'From the time of the ravages of the Black Death in Norway (1349) through the eighteenth century, Iceland was culturally isolated to a degree unknown in any other Western society.' Never subject to invasions, Iceland until relatively recently represented a petri dish of cultural conservatism, its institutions, laws, language and architecture all speaking of its isolation. Eiderdown harvesting seemed to be another sign of the Icelanders' unique position, their deep attachment to land and traditions that go back to the settlement period.

Just how special is shown by the example of another country with abundant numbers of eiders, Russia. Unlike Iceland, Russia has little history of a symbiotic relationship with the bird. Over centuries Russian eiders were slaughtered in the thousands, their nests plundered, their down plucked, possibly to fill the bedding of the Russian aristocracy. 'The unlucky creatures, ceaselessly harassed, flee their favoured islands, resettling in more tranquil territories,' wrote the Russian surgeon Franz Ul'rikh while based in Murmansk in 1877. 'They fly to the Norwegians.'

After the Russian Revolution of 1917, I imagined, eiderdown would have fallen from favour, sharing the fate of the class whom it had once warmed. Indelibly associated with excess and comfort, the material must have been tarnished, an unlikely filler for Lenin's duvet. Only later did the Soviets seek to find a place for eiderdown in their new universe, converting this symbol of excess into something utilitarian and nationalistic. 'They tried to sew uniforms for pilots, for astronauts, and for military use,' Jón Sveinsson told me, 'for uses other than luxury.'

During Soviet times, Russian biologists toyed with the idea of harvesting eiderdown in the Arctic on a collective basis. There were numerous plans to set up eider farms, but only one successful attempt to harvest down on a large scale. In the 1940s a remarkable young Russian biologist named Nina Demme managed to harvest more than five tonnes of eiderdown over five years in Novaya Zemlya, an archipelago in the Russian Arctic. An expert on eiders, she drew on the Icelandic method, setting up artificial shelters for the ducks, killing their predators, and enlisting locals to collect down. She stopped her experiment in 1945, less than a decade before the Soviets started using Novaya Zemlya as a test site for nuclear weapons.

If Iceland's eider colonies speak of its steady rhythms, its insulation from cataclysms, then Russian eiderdown is a portrait of the country's constant political upheavals, a series of stops and starts. Alexandra Goryashko, a Russian biologist who has spent decades researching the relationship between Russians and eiders, told me that she was sceptical that Icelandic harvesting could ever work in Russia. In the failed or short-lived attempts to harvest eiderdown there she saw not isolated incidents, but evidence of a pattern, the surface manifestations of Russia's ingrained political instability.

'The Eider is a conservative bird,' she wrote, 'and successful cooperation with it is possible only in a conservative society, with settled
traditions, in a society free of social cataclysms. Unfortunately,
there is no such society either in the USSR or in the present-day
Russia.'

And yet there were signs, too, that perhaps Russia's relationship
with eiders might become the dominant one, as Iceland continues
to be shaken by the effects of the financial crisis, the influx of millions of tourists and rural depopulation. Like the sheep farmers, the
eiderdown harvesters were getting fed up with the harsh weather
and isolation. 'The composition of the harvester is changing and
they are getting fewer and older,' Jón told me. 'In our case, we com-
pensate by getting foreigners to assist harvesting.' Many eiderdown
farms, he said, are merely summer houses, places of recreation for
city dwellers. 'The new owners are not interested in the pittance a
few kilograms of eiderdown give. After all, they have come to the
countryside to have a beer on the porch, fire up the grill, and relax
on the weekend, not to run around harvesting foul-smelling, flea-
ridden eiderdown.'

I tried to imagine what the Westfjords might look like in fifty
years or so, but I did not have to exert myself too much to envisage
one possible outcome. During a short break from talking with farmers, I took a boat out to the Hornstrandir, the desolate northernmost
finger of the Westfjords, which is cut off from the rest of the peninsula by a huge glacier. Once inhabited by whaling communities and
eiderdown farmers, it is now deserted except for the summer months,
a place of skittish eiders, huge populations of Arctic foxes and the
occasional polar bear ferried from Greenland by rapidly melting sea
ice. I spent four days there, walking alone on its barren east coast,
cutting across its deep valleys, passing bleached white driftwood

from Siberia, old rusted mines from World War Two, and a grave-yard whose most recent tenancy dated to 1949.

When I reached Furufjörður, a bay on the peninsula's eastern side, I saw a hot-air balloon stranded in the shallows in the distance, bobbing gently in the surf. I headed toward the rubber, but there was no basket: the body of a young blue whale lay wrecked in the sand, its tongue inflated with gas. It had, I later learned, died several days earlier. Already the stench of rotting flesh and fish pervaded the bay, its water scummed with white fatty deposits that pasted my walking shoes and trousers. Although swollen, the whale was strangely intact. No one had cut up its flesh, no one had claimed it, but I saw that I was not the first to have found it. Peering closely at its tail, already half buried in the sand, I saw that a small square of flesh had been removed, a symbolic gesture to old times, a sign of luck. Within several months, little would be left, its ribs protruding from the sand like the remains of a shipwreck. Perhaps Icelandic eiderdown would be no different, blowing away in the wind while the ducks pursued the Icelanders all the way to Reykjavík.

'It's not sad,' Jón told me, 'just different. The world is getting smaller.' But I did feel sad, thinking that this tradition might disappear. When I returned to the UK, I made a habit of checking in department stores to see if they had eiderdown quilts or pillows. Finding them in a shop window, I always felt reassured that the traffic in down continued, that this tradition still had a place in our own century of synthetics and factory farming. Nestled in silk covers, the down was always hidden from view, its rich history condensed into a short description on the label. Tempted by an act of mischief, I almost wanted to rip them apart, allowing the down and its stories to expand and escape outward.

EDIBLE
BIRDS' NESTS

THE BLACK-NEST SWIFTLET

I n June 1910, the British explorers Henry Bowers, Bill Wilson and Apsley Cherry-Garrard embarked on what the latter called the 'weirdest bird's nesting expedition that has ever been or ever will be'. Lugging two nine-foot sledges, they set off in pitch darkness from their base at Camp Evans in the Antarctic, hoping to cross the Ross Ice Shelf and reach Cape Crozier, where they planned to retrieve eggs from a colony of emperor penguins. Their journey, carried out in complete darkness, took thirty-five days, during which they braved blizzards, crevasses and temperatures as low as minus 60 degrees Celcius. On returning to the camp, cradling three eggs, they were practically unrecognisable after more than a month in one of the most hostile places on earth. Cherry, writes his biographer Sara Wheeler, 'looked about thirty years older than he had when he had set off, his cadaverous face scarred and corrugated, nose dark, eyes dull and hands white and wrinkled with damp.'

Of the three, Cherry was the only one to make it home to England; his companions later froze to death in their sleeping bags alongside Robert Falcon Scott during his failed attempt to be the first to reach the South Pole. Cherry remained back at base camp and was spared their fate; his later account of the expedition, *The Worst Journey in the World*, became a rallying call for all adventurers willing to risk their lives in pursuit of a lofty scientific goal. 'We are a nation of shopkeepers,' he wrote, 'and no shopkeeper will look at research which does not promise him a financial return within a year. And so you will sledge nearly alone, but those with whom you sledge will not be shopkeepers: that is worth a good deal. If you march your Winter Journeys you will have your reward, so long as all you want is a penguin's egg.'

In Victorian times the object from the wild often performed an important literary function in a specific popular genre, the quest romance. According to the critic Robert Fraser, its roots lie in ancient myths where a protagonist, usually male, sets off with a team of specially chosen companions to recover a prized object such as a 'golden fleece, or the skin of some fabulous animal.' In such stories it was crucial that the object be as strange as possible, shedding light on some essential truth. In this light, Cherry's choice of object, the penguin's egg, was masterly. As emperor penguins were 'probably the most primitive bird in existence', the retrieval of their embryo, he surmised, 'may prove the missing link between birds and the reptiles from which birds have sprung.'

I thought back to my own first object, the down of an abundant duck. No Victorian adventurers set off to retrieve eiderdown, nor is it mentioned by the authors of the quest genre: Haggard, Conan Doyle, Stevenson and Kipling. Perhaps down was just too familiar to readers, more the stuff of upmarket shopkeepers than of adventurers seeking the missing link. When I first held eiderdown in Iceland, I was spellbound by its lightness and coherence, but I could easily conjure its source in my mind, the breast of a duck or goose, bobbing somewhere in the Arctic Circle. The same could not be said for the next object that I found. In the basement of the Penn Museum in Philadelphia, the keeper of its Oceanian collections opened a box on a scrubbed white table as clean as an operating room. Inside were some thirty nests, chaotic half cups, seemingly fashioned of small black feathers and a brittle glue-like substance that varied from white to orange to red. The feathers appeared secondary to the glue, a sort of addition based on artistic whimsy.

Admiring these nests, I felt a sense of fragmentation: those miniature black feathers spoke of birds, but the encasing strands reminded me of sea creatures, the threaded tails of jellyfish, or the silky strands of byssus, those fibres emitted by mussels to root themselves to the sea floor. Such marine images intrigued early naturalists as they sought to deduce the building material of these nests. Like forensic investigators, they endlessly tested their properties, burning the strands, adding ammonia to them and later examining them under a microscope. 'They seemed', wrote an eighteenth-century English botanist, 'to be composed of fine filaments, cemented together by a transparent viscous matter, not unlike what is left by the foam of the sea, upon stones alternately covered by the tide, or those gelatinous animal substances found floating on every coast.' Others speculated that they were built of coagulated seashells, agarwood resin, sea plants of the genus *Gelidium*, oysters, jellyfish, 'marine polypi' or tissue, isinglass – the dried swim bladder of a sturgeon – or sperm from whales or fishes.

The nests form part of the collection of William Henry Furness III, a Philadelphia-born physician. In May 1896, at the age of twenty-nine, he and one of his university classmates, Hiram Hiller, travelled to the state of Sarawak in Borneo. The ostensible reason for Furness's journey was scientific. With the financial backing of his father, the Shakespearean scholar Horace Howard Furness, he intended to gather specimens and objects for the Penn Museum, then a fledgling institution. But he and his companions were driven by a thirst for adventure and a desire to make a name for themselves. 'They intend making a journey across Borneo,' reported a Honolulu newspaper. 'As much of the country has not been visited by a white man the trip is fraught with considerable danger. . . . They will have a siege of rough life such as they have never experienced before.'

In the minds of many young American and British travellers of the nineteenth century, the state of Sarawak was synonymous with romance, adventure and danger. About the size of England, it lay on the northern coast of Borneo in the heart of the Malay Archipelago. At the time of Furness's trip, it was ruled by Charles Brooke, a Somerset-born former officer in the Royal Navy, who oversaw his dominion from his palace in Sarawak's capital, Kuching. Like the heir to a family estate, Charles had inherited Sarawak and its peoples from his uncle James, who – in the tradition of colonial adventurers – resigned his commission in the British East India Company, bought a schooner, assisted the sultan of Brunei in repressing an uprising, and in 1841 was rewarded with the governorship of Sarawak, becoming the first 'White Rajah'.

Drawn by the promise of a civilising mission, young men lined up to wear the uniforms of the Brooke civil service, adjudicate tribal disputes, collect taxes on behalf of the rajah, and explore Borneo's interior. By the late nineteenth century, Sarawak's coast was well trod, inhabited by indigenous Dayaks and Malays together with Chinese, Arab, Indian and European traders. But its inner territory remained largely unexplored, the legacy of James Brooke's light-touch rule. 'Inland', wrote Charles Hose, a nineteenth-century Brooke officer, 'there are various tribes indigenous to the soil, barbarians in name and with many savage instincts.' Despite the efforts of the Brookes, headhunting was still practised by the inland peoples, their longhouses decorated with the fleshed skulls of their enemies.

Aside from his medical knowledge, Furness had few qualifications to enter into this world. He knew little about Sarawak's peoples, their diverse languages or culture, or indeed Borneo's extraordinary biodiversity, which had first drawn the great Victorian naturalist Alfred Russel Wallace. The young men arrived in Kuching in May 1896. Conscious, no doubt, of the expectations of Philadelphia's lecture circuit, Furness sought out the most exotic sides of tribal life in Sarawak. Guided by a Brooke officer, he and his companions journeyed along Sarawak's rivers, visited longhouses and recorded details about local life, from tattooing to headhunting practices. As the men travelled, they accumulated an astounding array of specimens and ethnographic artefacts, many of which would later prove useful in illustrating their lectures to packed halls in America: shields, blowpipes, medicines, poisons, insects, fish, orangutan skeletons and these birds' nests.

Furness collected his nests towards the end of his visit to Borneo. In August 1896 he left Hiller and travelled along the banks of the Dapoi, a tributary of the Baram, towards the Great Cave of Niah. One of the largest in Asia, the Great Cave had occupied a special place in the Victorian imagination ever since Wallace suggested that a Borneo cave might hold evidence of the 'missing link' between humans and apes. Almost ten hectares in plan area, its great arched caverns are often compared with a cathedral, but perhaps something is lost in equating its vast space with the chiseled angles of Notre-Dame or St Paul's. Countless passages, domes and crevices bewilder the visitor, monuments to the creative power of water, heaving with small birds and bats. 'Insensate, indeed,' wrote Furness, 'must be he who is not filled with speechless awe as he turns

from the brilliant sunshine and enters this illimitable abode of
silence and of night.'

Although Niah was a short distance from Sarawak's coast, its
limestone caves were cut off by a sea of uncultivated forest, home to
orangutans, gibbons, snakes, and myriad birds of all shapes and
sizes. 'It is the ideal forest primeval,' enthused Furness, 'where old
and majestic trees form in every direction vast, illimitable, solemn
vistas.' Travelling through this 'untouched land', he encountered the
Penan people, hunter-gatherers who wore loincloths and lived off the
forest, hunting their quarry – monkeys, snakes and porcupines – with
blowpipes and poison-tipped darts. 'Gentle, simple-hearted creatures, they believed that merely to stroke us or to be stroked by us,
brought them blessings. . . . They examined and admired everything we had with us or on us; our coats, our hats, our shoes, the
buttons and buttonholes on our clothes, – these excited their profound wonder.'

When Furness approached the cave, he described it as a place of
death and darkness. 'It seemed the veritable entrance of the Inferno,'
he wrote. 'As the light from the opening struck the massive projections here and there, and cast long, blacker shadows, it became a
landscape in the moon, while the appalling, death-like stillness
seemed to presage a frightful cataclysm in nature.' The cave seemed
as unknowable as the surface of another planet, a black hole or tunnel that led only to death and decay. 'The extent beyond, in the utter
darkness, seemed illimitable.'

The root of the word cave – cavus, or 'hollow' – is highly misleading. In the dark, humid conditions of caves, an abundance of creatures have evolved that continue to confound naturalists: the olm,
an aquatic salamander, first described as a baby dragon in the seventeenth century; the blind cave crayfish, which was reported to live

for 175 years; reticulate pythons, which can negotiate steep walls to find prey. The caves around Niah were no exception: they had the largest colony of naked bats in Sarawak, the rare earless monitor lizards, and, as discovered later, a particular species of gecko. When Furness entered, he triggered a great exodus of small black birds that behaved like insects: 'Our presence and the echoing of our voices soon startled the swallows, and forth they emerged, in myriads on myriads, from the darkness, and circled round us and above us, and about the mouth of the cave like swarming bees; the whirr of their wings and their twittering sounded like waves on a pebbly beach.'

The birds that Furness had disturbed were in fact not swallows but the black-nest swiftlet, *Aerodramus maximus*. A small swift, it roosts and nests deep in caves throughout South East Asia, leaving only to feed on insects. Unlike most swifts, it has special adaptations for life in caves: tiny feet that allow it to cling to limestone and large black eyes that can make the most of low-light conditions. Though Furness did not know it, the sound he heard in the cave, the 'waves on a pebbly beach', was the noise of the birds' echolocating clicks, an adaptation that allows them to navigate in the cave's total darkness, not unlike the sonar of bats, but at much lower frequencies within the range of human hearing. The birds make these clicks by contracting muscles on both sides of their syrinx, an enlargement at the lower end of their trachea where the two bronchi merge.

Like their feet, eyes and clicks, the birds' nests, gummed against the cave wall, mark them out as cave dwellers. On entering a cave on an islet near Java, a French missionary wrote, he 'perceived the roof of the cavern to be covered entirely with small nests shaped like holy-water pots.' Although many surmised that these nests were built of a marine substance, in 1817, Sir Everard Home, a British surgeon, cut open a 'Java swallow', examined its glands and put such theories to rest: 'There is a membranous tube surrounding the duct of each of the gastric glands, which, after projecting into the gullet for a little way, splits into separate portions like the petals of a flower. That the mucus . . . is secreted from the surface of these membranous tubes, there is no more doubt.' In other words, the birds themselves produce the substance of which the nest is crafted.

During breeding season, both male and female birds begin to retch and chew, excreting small strands of a thick, gelatinous substance from these modified salivary glands lying below their tongue. This they spread in arched form across the cave wall, inserting dark brown or black feathers from their plumage. After thirty days, the initial arch has grown to form a shallow cup into which the female bird lays one egg. Other kinds of swifts rely on saliva to build their nests, gluing twigs, vegetation and other materials to form cups or loose balls. But examining these nests in the museum, I was struck that the black-nest swiftlet crafts its nest *entirely* from itself, using only its saliva and feathers. I craned my neck over the nests to try to smell any hint of the cave, but my enthusiasm was dented by the keeper's warning that the collection might contain traces of arsenic, which Furness had taken with him to preserve specimens.

In his account of his travels, Furness does not pause to reflect on the nests he collects, their gummy material, their essential mystery and strangeness, except to say that they exercised a certain power over the Penan, drawing them away from the forest. Like rhino horn, beeswax and camphor, the nests were so valued by Chinese traders that these forest dwellers would set aside their foraging ways to obtain them, ascending high wooden poles set deep into the cave. 'Up these poles', wrote Furness, 'the agile [Penan] climb hand over hand and foot over foot, walking up them like monkeys. When at the top, they scrape down the nests within reach, by means of a long pole bearing a hoelike blade, and with a home-made wax candle fastened to it to show where the nests are.'

It is not known exactly when the Chinese first started trading in – and eating – swiftlets' nests. It has been suggested that the nest trade may go back as far as the T'ang (618–907) or Sung (960–1279) dynasty, though the first actual reference to the nests in Chinese literature appears later, in the early Ming period (1368–1644). Despite the paucity of historical sources, stories abound as to the origins of the trade. Many in China believe, though not on the basis of any documented evidence, that the fourteenth/fifteenth-century Chinese admiral Zheng He brought back birds' nests from South East Asia as a gift for the emperor. Contemporary authors, including one of the admiral's scribes, do not mention nests, but the trade in swiftlet saliva is now tied – at least in the imagination – to the great era of Chinese naval expansion, when the Ming empire sent out fleets to Asia and East Africa.

Whatever truth lies in these stories, from the seventeenth century the nests became a major export commodity, appearing in lists of royal tributes from South East Asia, popular novels, trade registers and imperial banquets. Associated as they were with emperors and decadence, there was no object that could evoke wealth and prestige more than swiftlets' nests. But they purportedly had magical healing properties, too, much like the metal elixirs once ingested by Chinese emperors. Traditionally boiled in a broth, the nests are or were believed by many in China and Hong Kong to reinforce bodily fluids, nourish blood, aid metabolism, treat cancers and AIDS, and refresh the 'weary frames' of opium smokers. 'In short', concludes Leonard Blussé, a Dutch historian, 'birds' nests were thought to be a cure for almost anything.'

When we take an object from the natural world, we often change its form but not its function. Nestled in a duvet, eiderdown provides us with warmth; a glass of milk offers us nutrition. But eat a nest and you transform its original function, turning a bird's home into a foodstuff. 'Anything *less* like good food is hard to imagine,' wrote the ornithologist and amateur archaeologist Tom Harrisson in his 1960 essay 'Birds and Men in Borneo'. 'These small, hard, usually fouled crescentic cups would seem to be more suitable for making furniture than soup.' American and European scientists have found little in the way of special nutritional value in a nest: one food scientist established that the saliva of a nest contains about 60 per cent protein, 25 per cent carbohydrate and 10 per cent water, with the remainder composed of inorganic ash. 'I am . . . of the opinion', commented Francesco Redi, a seventeenth-century Italian physician, 'that we owe this ingenious invention solely to the epicureanism of the later ages, which, always, hungering after novelty, sets an adventitious value upon what is brought from a distance and difficult to be procured.'

In 1922, Albert M. Reese, an American biologist, took it upon himself to taste a range of 'animal foods not commonly eaten by Americans.' In the name of science, he ate monkey ('very agreeable'), bush-hog ('tough, but is excellent'), opossum ('quite pleasant'), groundhog ('gamy'), muskrats (similar to squirrel and rabbit), and salamander eggs ('very pleasant flavor'). One specimen in particular disappointed him: the nest of the swiftlet. Reese wrote: 'Properly cooked the edible bird's nest is supposed to be a remarkable delicacy. The writer prepared, according to the only recipe available, a nest obtained near the island of Palawan of the Philippine group. The result was a gelatinous mass without a particle of taste. Either the nest was stale or there was something wrong with the method of preparation, since none of those who tasted it was enthusiastically anxious for more.'

In the writings of many historians, scientists and colonial administrators, the bird's nest came to symbolise all that was different between the 'Orient' and the 'West', proof of a gap that could not be breached. If eiderdown created utopian visions among travellers and writers, then the trade in birds' nests often resulted in estrangement, a sense of distance. Furness's descriptions are no exception to this. When he writes of birds' nests, he is *bewildered*, literally 'lured into the wild', describing the infernal darkness of the cave and apelike men ascending wooden poles at the behest of Chinese nest eaters. Years later he would write that the Penan were 'the most primitive type of human being inhabiting the island', whose very nature led him to question the size of the 'step' between humans and apes, such as the orangutan and chimpanzee. Inspired by what he saw in Borneo, he returned to Philadelphia and spent years trying to teach orangutans how to speak from his home in Wallingford.

Today I notice that little has changed. Read American or British press articles, and it is hard not to be bewildered by the trade in 'bird spit', the sight of muscled harvesters risking their lives to retrieve nests deep within caves. While the United States had exported billions of dollars' worth of pharmaceuticals to China, there was only a trickle of swiftlet saliva from across the Pacific. One could try to make sense of the different facets of Chinese life, from the workings of its Communist Party to its complex characters to its rich ceramic tradition, but the bird's nest somehow seemed impossible to decode, an enduring symbol of difference. As I read the news of trade wars, retaliatory tariffs, accusations of currency manipulation and tensions in the South China Sea, I feared that the bird's nest was somehow acting as a silent prop in these conflicts, proof that two distinct civilisations were destined to clash.

After they returned to Philadelphia, Furness and his companions became regulars on the city's lecture circuit, the American Philosophical Society and the Geographical Society of Philadelphia. With its descriptions of the lives and tribal practices of hunter-gatherers, and images of bare-chested women and caves, Furness's account of his travels, *The Home-Life of Borneo Head-Hunters*, was a great success in the United States, temporarily assuring his status as one of the country's foremost adventurers and interpreters of Borneo. But I felt dissatisfied by his account of birds' nests. Like a penguin's egg, the nest of a swiftlet seemed to serve as an auto-biographical prop, evoking a sense of strangeness and danger in Borneo that was concomitant to his own courage and boldness. Instead of combining the stories of the harvesters, traders and nest eaters, he'd plucked the nests en route to China, cutting their journey short, ossifying them to serve a single narrative purpose. I wondered where these nests might have ended up had they made it to China, crossing borders, exchanging hands and entering gullets. Was it possible to reconfigure the nest, to turn it from something alien to something intelligible? To overcome the gap between this 'fouled crescentic cup' and a prized delicacy, the 'caviar of the East'?

One summer I travelled to Sarawak, first flying to the city of Miri, the steaming heart of Malaysia's oil industry. In contrast to the situation in Furness's time, it required little in the way of planning, logistics or nerve to visit Niah – merely a contact in Sarawak's Forest Department, Haidar Ali. Haidar shares his name with a military commander from southern India who resisted the British East India Company, but he was a gentle, quiet man who had spent most of his career studying Sarawak's hornbills, swiftlets and geckos. He'd recently left his job as the warden of Niah National Park, overseeing its numerous caves and population of swiftlets, and he seemed relieved to have escaped the responsibility.

A coastal road now connects Miri to Niah, a direct line of tarmac stretching close to the caves' entrance, rather than the winding rivers that once proved the fastest means to penetrate Borneo's interior. Sitting next to Haidar in his car, I thought of Furness's descriptions of Sarawak's 'primeval' forests. From 1963 to 1985, it is estimated, 30 per cent of Sarawak's forest was logged. Faced with falling timber prices, the Sarawak government instead chose oil palm, turning vast tracts of land into monoculture. Driving along the coastal road, we passed through a new forest of oil palm plantations, the even spaces of palms replacing the varied hues of primary rain forest. I didn't see any nomadic hunters, loincloths, or blowpipes, only oil palm trucks pouring out of the town of Batu Niah, Indonesian labourers from Kalimantan, and a handful of Chinese traders, many of whom settled in the area during the timber boom in the 1970s.

I'd headed to Niah to meet one of the region's most famous birds' nest harvesters, Nuar bin Haji Jaya. Half Penan, half Malay, he is descended from the harvesters described by Furness in the Great Cave. An expert climber, Nuar has spent his life working in Sarawak's limestone caves, earning the nickname Spider-Man. Unlike the rock climbers I knew, though, Nuar was not a lean, muscled figure, unencumbered by a single gram of fat; when I arrived at the national park, I found a stocky figure, his paunch pressing against his black sports shirt. He carried a small black bag whose strap, pulled tightly across his chest, followed the rolling contours of his stomach and chest. When he walked, he lumbered.

After some preliminaries in Malay, he invited me to his cave, Lubang Perintah, which lay a twenty-minute walk through the forest. As we wandered through it, my gaze was drawn to the wooden hand-rail of the walkway, covered in beautiful miniature creatures: hairy caterpillars, bright white bugs, orange millipedes and hammerhead worms. In the late 1950s, Tom Harrisson eloquently described the noises of Borneo's forests: 'tree crash, cicada buzz, mongoose chuckle, the whistle of the blood-red and black hill partridge, grasshoppers, a million moving termites, piglets, bat swing, goat laugh, eagle owl.' But all I could hear on the way to the cave were Nuar's roars of laugh-ter as he recounted his feats of daring, from climbing in caves to swimming in Sarawak's crocodile-infested rivers. Fearless and bra-zen, he told me that his motto in life was a common Malay expres-sion: 'When faced with a tiger, you should roar.'

Eventually we came to a limestone cliff, halfway up which lay a wooden hut, at the entrance to a large cave. It was reached via a rickety wooden staircase that rose some thirty metres from the forest floor. Shortly after we arrived a storm broke out, and we instinctively headed for Nuar's hut, within the cave mouth. Looking outward, I spotted a hornbill, its huge wings beating over the forest. How tempting it was to imagine this place as wildness itself, but the illu-sion was broken by the low roar of the oil palm trucks and the smoky dust of a quarry, eating away at the limestone in the distance. I felt safe in the cave, untouched by the rain as it thrashed the canopy below, but I soon learned that we were subject to another precipita-tion, the guano of birds and bats. Soft, moist and peaty, it formed thick beds around the hut, which vibrated with insects and dying birds. When I walked on these guano beds, they swallowed my boots, which left deep imprints in the brown snow. The guano covered everything, blackening my hands and the white linen shirt that I had inadvisedly chosen to wear.

A more insalubrious place was hard to imagine, and yet Nuar had turned the cave into a domestic space. He wandered around the hut in a sarong, his chest bare, singing to himself and watching videos on his phone. Resigned to the guano, he could wash in a makeshift shower in a side chamber inhabited by a family of bats. It was an ingenious contraption, fed by drips of water from a stalactite collected by a large plastic sheet that led to a tube. Only the excrement's inhabitants riled him, countless crickets, flies, millipedes and ants. To thwart them, he stored his food on a large shelf under his table, its legs placed in tin cans full of water; he waged war with a can of fly spray and a tea towel.

Historically, tropical caves have been considered dangerous places, hostile to our own species. In 1956 the German paleontologist Ralph von Koenigswald dismissed the idea that early tropical humans had any interest in 'damp caves inhabited by snakes, bats and evil spirits.' But later that same decade, Harrisson conceived Niah's caves in radically different terms. Inspired by Wallace's writings, he theorised that early man had lived in the Great Cave, feeding off its population of swiftlets and bats. 'If there ever were early cave men,' he asked, 'where better to live than here?' Unlike other scholars, he and his second wife, the archaeologist and art historian Barbara Harrisson, described Niah as a 'larder', offering 'relief from want, refuge from elements, and liberation from society.'

Watching Nuar, I could not help thinking of Harrisson's view of the cave, his conception of its caverns as places of comfort. Walking in his flip-flops on the guano, the harvester seemed so at ease in this space, liberated from the demands of his family or the downpours of Borneo's monsoons. And yet it was not the prospect of nutrition – the flesh of bats or swiftlets – that drew him to this dark space, but the nests' exchange value. By selling the nests to traders, he had grown fantastically rich over the years, investing in a small oil palm plantation, taking a second wife, and employing the services of a manservant to do his cooking and cleaning in the cave. However comfortable he was here, the cave was a place of work, a source of wealth rather than subsistence.

Keen to show me the source of this wealth, Nuar led me into the bowels of the cave, beckoning me across the guano beds. As we proceeded, the green moss on the limestone grew darker and then disappeared. Every so often we would pass a baby bird stuck in the guano, flailing to rid itself of the sticky excrement, but only digging itself deeper. Above us we heard an electric clicking, the sound of a thousand muffled typewriters. Nuar walked on towards the sound, climbing up a slippery limestone tongue, and called my name. Pointing his torch beam above us, he gestured to the rows upon rows of nests high up on the cave wall. Within them, the swiftlets crammed, their small wings forced upwards. Nestled in grooves and crevices, the nests almost seemed outgrowths of the limestone itself, black and white cups gummed against the pallid greys of calcium carbonate.

Between 1854 and 1856, Alfred Russel Wallace travelled through the south-west of modern Sarawak, gathering specimens, shooting orangutans and formulating his theory of natural selection. He never reached Niah, only recommending the excavation of Borneo's caves to look for remains of ancient hominids, but I always wondered what this brilliant observer would have made of swiftlets' nests. In writing on his travels in Borneo, Wallace delighted in the durian, a large sweet fruit that was prized for its delicious flesh and grew high on tall trees. He was struck that a fruit so valuable should be so difficult to harvest, its hard shells often falling from trees and killing anyone unfortunate enough to be below them. 'Trees and fruits,' he quipped, 'no less than the varied productions of the animal kingdom, do not appear to be organised with exclusive reference to the use and convenience of man.'

Perhaps Wallace would have spoken of birds' nests in a similar fashion. Like the durian, the nests were prized objects, sought for their special qualities, but they were gummed high up within the cave, well beyond the reach of cave snakes, the large carnivorous insects that prowled the guano, or the hands of harvesters. Nuar could reach them only by relying on a network of bamboo poles and wooden ladders that he had built within the cave. Wedged high in the limestone, they resembled matchsticks or fish bones, the work of a carpenter unhinged. Peering towards them, I felt dizzy, my own internal spirit level upset by the varied angles of the wooden spars that spliced the cave's space. I recalled the wooden poles I had once seen within Serbariu, a vast abandoned coal mine in Sardinia, which propped up the mine's roof, and I was overcome by a terrifying sense that if Nuar's handiwork were removed, the entire edifice of the cave would collapse.

Nuar offered to take me up, a large grin on his face, but I was not too sure. Even with decades of experience, it is extraordinarily dangerous to collect birds' nests. 'Every now and then someone falls,' remarked Harrisson unsentimentally. 'To such a fall there is only one answer short of a miracle.' Perhaps unsurprisingly, in recent decades the birds' nest harvest has become a staple of television documentaries. In 2011 the TV chef Gordon Ramsay climbed deep into a cave to retrieve a bird's nest in Sabah, Malaysia. 'I've gone to great extremes to get some fantastic produce before,' he tells his guide, 'but I'll tell you now: I hope this tastes fucking amazing.' Before I came to Niah, I watched a fair share of these programmes, but standing in the cave, I appreciated that any attempt to photograph or film the birds' nest harvest would inevitably fail to capture its essence. Much of Nuar's harvest takes place in absolute darkness; by lighting the cave, one inevitably diminishes its peril.

In 1956, Tom Harrisson and the filmmaker Hugh Gibb set out to document the harvest of birds' nests at Niah for an episode of a BBC television series, *The Borneo Story*. Backed by the BBC, the pair travelled to the Great Cave with two generators and electric lights to illuminate the cave for the first time. Long before the disappearance of Sarawak's forests, it was a complex logistical operation requiring river transport, porters and extensive planning. '[Harrisson] ran the operation like a military expedition,' recalled Sir David Attenborough, who later edited the film in London. 'They had to sail up the coast from Kuching, then continue upriver, hack their way through jungle, build a walkway, and then shift all their kit, including some huge generators, into the caves.'

In the 1975 edition of *Who's Who*, Harrisson described his chief recreation as 'living among strange people and listening to them talk about themselves.' By the time he filmed birds' nests, he was embarking on his fifth or sixth career, having previously made a name for himself as an ornithologist, explorer, radio critic, anthropologist and soldier. He first came to national attention at the age of nineteen when he organised a nationwide survey of the great crested grebe in Britain. After dropping out of Cambridge, he joined various expeditions to the New Hebrides and Borneo, then founded Mass Observation, a pioneering project to document life in British towns, in 1937.

In Harrisson and Gibb's episode, titled 'Birds' Nest Soup', the camera follows a party of muscled birds' nest harvesters entering the cave, bearing long poles of ironwood. To these men the cave seems a familiar space, and they move with the ease of commuters, entering the mouth just as a group of guano collectors have finished the night shift, their backs laden with the black soil to be sold as fertiliser. There are no references to the Inferno or death-like stillness; the cave is alive, a vast pulsating organ, breathing birds and bats in equal measure. We see the swiftlets, clinging to a cave wall, their miniature feet like hooks; we see a close-up of the wings of a bat extended by Harrisson's assistant, his fingers dwarfed by those of our distant mammalian cousin. These two creatures are different and yet alike, finding similar solutions to the caves' restrictions, its angles and lack of light. We hear the sounds of the cave, its squeaks, clicks and squeals reverberating through the darkness, and can almost smell the reek of guano.

Within the cave, the harvesters erect a mast, lashing it to stakes in the guano with rattan ropes. The camera lens follows a harvester, Johari, ascending an ironwood pole hundreds of feet into the cave's heights. As the harvester climbs, it is as if one's vision had been rotated ninety degrees; his taut limbs proceed up the vertical pole as if he were walking on all fours along a flat surface. The climber's prehensile toes resemble another set of fingers, gripping the pole and propelling him upwards. Like Furness, Harrisson could have alluded to apes in this scene, but he chose to compare Johari to a circus acrobat. Although the act of pole climbing might evoke our ancient tree-dwelling ancestors, it is one that is carried out with the use of arched feet with non-opposable big toes, feet that are no different from those of the shoe-wearing BBC viewer he addresses.

Harrisson's first major book, *Savage Civilization*, published in the 1930s, detailed his experiences among the cannibals of the New Hebrides. It is a mash of personal observation, history and travel narrative, but above all it seeks to undermine conventional divisions between savagery and civilisation, dwelling on colonial and missionary injustice and the complexity of the societies of 'so-called primitive peoples'. In its own way, 'Birds' Nest Soup' does the same. Unlike Furness, who writes of harvesters as monkeys, Harrisson conceives of the harvest as more than a physical endeavour of muscle and sheer physical bravery. He shows us dense ironwood poles, rattan ropes that hold the poles in place, and long bamboo scrapers assembled in sections, which slip into one another as smoothly as moulded metal; he shows us that the entire enterprise of harvesting depends on technology and divided labour – in other words, culture.

At the top of the mast, hundreds of feet above the cave floor, the collector enters an enormous limestone dome, blackened by nesting swiftlets. Great swarms of birds swirl around the harvester, darkening the pale tones of the dome. Poised on the mast, the harvester uses a scraper attached to a bamboo pole to remove the nests gummed to the cave wall. Like harvested fruit, they rain down onto the cave floor, to be collected by his assistant. As if these feats were not enough, Harrisson refers to the darkness of Niah, its 'twenty-four-hour-long night', fully aware that he has converted a private act of labour into a public display of acrobatics.

After the harvest, Harrisson shows us the transition of nest to foodstuff, moving from the depths of the cave to the Niah River, which I had crossed to get to Nuar's cave. Nearby, a young Chinese woman soaks the black nests, their strands swelling to form gelatinous clumps double the original size. She fastidiously picks out the feathers and dirt from the nests, transforming them into translucent flat pancakes that dry in the searing sun. Once they are dried, a Chinese trader weighs a great pile of flat nests on a scale before Harrisson transports us to a kitchen, where a plucked chicken is stuffed with the gelatinous strands of swiftlet saliva and boiled in oil. In one of the final scenes, a group of Chinese diners, elegantly attired in evening dress, help themselves to bowls of soup on a spotless white tablecloth. 'The Chinese, like any other people, love anything that's strange and exciting to eat,' Harrisson explains.

It intrigued me that Harrisson could take an object like an edible bird's nest and use it to question differences between cultures, bridging the living room of a BBC viewer and the depths of the Great Cave. But when I thought about the metaphors commonly used to describe the collection of nests – the harvester as mountaineer or acrobat – they only reminded me of the gaps, the differences between climbing for birds' nests and alpinism or circus performance. Rather than conquer a summit or entertain, the climbers in Harrisson's film ascend to retrieve nests, nests that will ultimately become the subject of his work, *soup*. I struggled to imagine placing one in my mouth, my own saliva mixing with that of a swiftlet.

I appreciated Harrisson's film, but thinking back to it I now feel uneasy. It was sometimes said that Harrisson was never able to shed his cap as an ornithologist, viewing men largely as he viewed birds. In 'Birds' Nest Soup' all of the bodies are Malay or Chinese, but we hear only from one powerful narratorial voice, Harrisson's own. His film moves seamlessly among the caves, processors and a restaurant, tracing the passage of the nests in a single breath. Perhaps my own work was no different. As I spoke no Malay, I relied on a young translator to talk with Nuar, harvesting fragments of his life and sticking them together like film reels. Nor did I ever ascend the poles myself, experiencing the collection of nests first hand. I had dreamed about ascending these structures, but when Nuar offered,

I shook my head, my mind bombarded with images of my own body tumbling down the limestone and thumping into the guano.

Before the invention of beehives, honey was gathered wild by harvesters, often relying on smoke to disperse the insects. In Valencia, Spain, there is a cave painting believed to be eight thousand years old that bears a remarkable resemblance to the nest harvest in Harrisson's film. It depicts a human figure scaling a ladder to raid honey, his or her form surrounded by swarming bees incensed by the destruction of their home. 'Clearly,' writes food historian Bee Wilson, 'honey was so precious that men would risk death to get it.' Like wild-honey harvesting, the scraping of swiftlet nests is inevitably destructive, depriving the birds of their home and sometimes their eggs. Such techniques make swiftlet populations in caves particularly vulnerable; take too many nests and the population of birds would simply collapse, their eggs spilled onto the cave floor.

Under the Brooke administration, an accord existed between the harvesters and the birds at Niah, a limit on the harvest, which was monitored by government officials. During certain seasons of the year the cave would be closed, its birds allowed to breed undisturbed. Despite his swagger in his hut, Nuar viewed the caves as sacred places, naming them after spirits who had first led him to their hidden entrances. This particular cave is government-owned, in any case, and Nuar's harvest is carefully overseen by the state of Sarawak. Shortly after Nuar discovered the cave, Lim Chan Koon, a Malaysian ornithologist and swiftlet specialist, carried out a detailed assessment to ensure that the harvest was sustainable. Under Dr Lim's guidance, the harvest of nests is strictly controlled: the cave is closed every year for four months and routinely cleared of rubbish; Spider-Man's bank account is even subject to scrutiny by government officials.

In Harrisson's essay 'Birds and Men in Borneo', he documents the wide range of avian-human relationships on the island, describing 'birds as omens, guides, indicators, culture heroes and symbols'. 'There is probably no other part of the world where birds and men are more intimately intermixed than in Borneo,' he writes. 'Here birds are interwoven into the whole texture of thought and belief.' Staying with Nuar, I came to see his relationship with swiftlets as part of this unique intertwining, a private dance between its inland peoples and swiftlets. Yet, as in Iceland, this tradition was born not of isolation, but of Sarawak's centuries-old integration within a trading system. Like the down harvesters, Nuar rarely made use of the material he harvested, snacking only on the occasional nest. ('That is why I have such good skin!' he told me.) The nests in his hut would be picked up by Chinese traders, processed, and then eaten by the 'men and women of ambition' on the Chinese mainland or in Hong Kong. With one exception. Before I left he handed me one of the nests he had scraped from the cave wall. I rummaged around in my rucksack to see what I could give him in return, but found only my second-rate torch. He examined it sceptically and uttered a phrase that roughly translates as 'It's the thought that counts'.

As the light in the cave began to fade around six, Nuar was always silent and never cooked any onions, for fear of disturbing the birds. Gradually the swiftlets gathered from their day's feed, their bodies filled with insects. First there were just a few tiny black projectiles, but their ranks soon swelled, forming clouds of black snowflakes. They entered the cave and began to stream through the ancient coral reefs in large shoals, tracing the old pathways of acidic water. So tightly did they group that it was impossible to make out their individual forms among the billows. Nuar sat in his favourite spot on a wooden bench facing out over the canopy, the best seat in the house, admiring their return.

When I arrived back at the park entrance the following morning, I collapsed in one of the empty cabins, exhausted after a sleepless night in Nuar's hut. Away from the cave, the heat was unbearable, the fan in my room circulating only the hot air. On waking, I took repeated cold showers to cool down, scrubbing myself violently to get rid of the brown gunk under my nails. Even after showering, I scratched myself continually, checking my armpits for any sign of insects I might have carried from the cave. Harrisson wrote of the cave as a place of comfort, but all I wanted to do was remove myself from its smells, itchiness and sounds, and return to Miri. And yet I knew that I could not leave Niah without visiting the Great Cave, the origin of Furness's nests and the site of Harrisson's film.

It is said that a Penan hunter, Melibong, rediscovered the Great Cave in the 1840s after its entrance had become covered by vegetation. According to local tradition, Melibong initially mistook the nests he found there for fungi. 'All over the cave walls', Harrisson wrote, 'he found what he thought were mushrooms.' Much like a gold prospector, a Malay trader took samples 'of this rediscovered treasure' to the sultan of Brunei, which ultimately led to the start of the birds' nest harvest at Niah. Since Melibong's rediscovery, the Great Cave has been one of the centres of the world's production of black nests, a vast factory, yielding tonnes annually, estimated for 1931 to be 70 per cent of the harvest in Sarawak. Shortly after

Harrisson made his film, he proposed that there were about a million pairs of swiftlets within the cave.

Almost every visitor to Niah whose account I had read came back transformed by the experience of entering the cave, as if it possessed great secrets. In October 1954, Harrisson and the archaeologist Michael Tweedie started to dig into its guano to find the remains of ancient humans, convinced that Niah acted as a sanctuary for our distant ancestors. Much of the archaeological establishment thought them mad, but in 1958 they found a human skull buried deep in the cave's guano at a level that suggested it was some forty thousand years old, then the oldest skull of any modern human discovered in South East Asia. Its discovery set off a firestorm of debate among archaeologists, not least because Harrisson disdained modern archaeological methods, but his dating was later proved to be roughly correct.

I planned to visit the cave early in the morning to catch the swiftlets as they left. Wanting to time my journey just right, I arranged with a boatman to take me across Niah's river, the only obstacle en route to the cave. After a dinner of Chinese seafood soup in Batu Niah, I slept soundly in my cabin, relieved to be away from Nuar's kingdom. The next morning I set off for the Great Cave at around five, having found the boatman waiting. It was pitch-dark, and my spare torch proved dimmer than the glow of the fireflies in the forest. Following the wooden boardwalk through the karst forest, I tried to make out the sounds, the buzzing of cicadas or the *wah wah* of a gibbon, but I was soon distracted by the sound of rainfall and the boom of distant thunder as the canopy lit up ahead of me. Even with the rain, I sweated profusely, feeling mildly afraid without knowing why.

I pushed onwards, my clothes sticking to my skin, and it was a relief to reach Gunung Subis, the huge limestone outcrop in the heart of the park, and take shelter. Following a raised wooden walkway, I passed into the Traders' Cave, an antechamber that once served as a village for harvesters, and listened to the rain outside. Shining my torch against the pale stone, I could make out the varied hues of lichens, fed by the presence of light from the cave's exposed side. I headed on into the Great Cave, entering through the West Mouth. Above me the air was thick with sound, the combined squeals and clicks of thousands of swiftlets and bats. I tried shining my weak light up towards them, but the effect was similar to that of directing a beam at the sky. Unable to see the colonies, I could only imagine their numbers in the limestone, their calls echoing through the domes above me. As I listened to them, different images shot into my mind: a Geiger counter, a typewriter, rainfall pattering on stone.

At the West Mouth, I sat down on the guano hardened by the passage of thousands of harvesters' feet. From my seat of guano, I could just about make out the site of Harrisson's excavations, deep trenches in the cave floor to my right. I thought of Harrisson's words, his conviction that this space provided comfort. I wanted to feel his attachment to the cave, but I felt like a foreign body here, impatient for light to enter its halls. In the darkness it was hard to know where the cave started and where it ended, its scale impossible to fathom except for the minuscule light of a distant guano harvester toiling in the cave's depths. When the light from his lamp began to grow larger, I felt reassured by the possibility of company; as he passed on a wooden walkway, finishing his shift, I greeted him a little too loudly, my voice sounding alien among the clicks.

I was relieved as the sun finally began to rise. Great shafts of light gradually entered the cave, revealing enormous stalactites, hanging like teeth from the cave mouth. I had expected the lime-stone to be white, corpse-like, but instead the cave was velveted with lichens, ranging from pale to intense dark green. Another visitor appeared at the West Mouth, a tourist rather than a digger, and my confidence grew. I walked farther into the cave, peering up towards the permanent structures of the birds' nest harvesters, great poles of ironwood that pierced the cave's centre. I knew them to be as tall as the trees from which they came, yet they appeared matchstick-size within the immense interior of the cave. Finding the poles even in its highest, most inaccessible parts, I was astounded to think that this place, so large and so dark, had been scaled, mapped, almost scaffolded.

I paused below the birds' nest structures and looked out towards the West Mouth to watch the swiftlets leave. I'd read about huge whirling rings of birds at Niah, how its chambers 'become thick with calling birds, milling in the hot twilight in their thousands'. I'd read how the cave was once a nest factory, churning out nests by the tonne. I'd read of the cave as one of the great wonders of the natural world, this vast pulsating organism that breathed birds in their thou-sands. But as the light entered, all I saw was a gradual trickle of birds leaving to feed, the sound of clicks slowly fading above me. It was as if the cave had been emptied.

At the time of Harrisson's documentary, birds' nests remained beyond the reach of most Chinese families. The most expen-sive products that a Chinese family could own were bicycles, wrist-watches and sewing machines. Under Mao, the consumption of birds' nests was said to be discouraged, an unacceptable symbol of wealth. But in the 1980s the very idea of birds' nests began to change, and with it, its place in Chinese life. In 1978, China began its transition from communism to state capitalism, favouring economic growth over ideology. 'It does not matter whether the cat is white or black', China's new president, Deng Xiaoping, was reported to have said. 'So long as it catches the mice, it is a good cat.' Living standards rapidly improved and a new middle class began to seek consumer goods: washing machines, televisions and birds' nests.

Historically, the value of birds' nests has always been high, often compared to that of silver. But in the 1980s the price of the nests began to rise at an unprecedented rate, one that rendered obsolete any such comparison. According to the ornithologists Lord Cranbrook (Gathorne Gathorne-Hardy) and Lim Chan Koon, by 1999 the retail price of a kilo of cleaned black nests in Hong Kong could reach $1,800, while a kilo of white nests, those of a different swiftlet species, could be more than $6,600. Cranbrook and Lim expressed little doubt as to the cause of this steep rise. 'Birds'-nest prices', they write, 'are a more useful barometer of economic welfare in Eastern Asia than any Hang Seng or other Stock Exchange!'

In the depths of the Great Cave, the logic of the market, terrible and efficient, played out. Spurred by high prices, hundreds of harvesters moved into the bowels, denuding the limestone walls and silencing the swiftlets' clicks. The accord, the limit on the harvest, which had been in place since Furness's visit, was torn up, replaced by the logic of the mine. First the accessible seams of nests disappeared, the low-hanging fruit that could be plucked with ease. As these strands dwindled, the price of nests rose even higher, pushing the harvesters into the most inaccessible and dangerous parts of the caves. The caves' traditional owners, the settled Penan people of Niah village, lost control of their seams, relinquishing their titles in a complex set of leases and subleases. Property rights, once secure, were up for grabs; disputes in the caves often ended in violence.

With the birds deprived of their nests, their population plummeted, their cracked eggs littering the guano. Lim Chan Koon told me that the Niah Cave suffered a decline of 80 to 96 per cent of its swiftlet population. The sudden passage of so many human hands and feet into the cave was notable not only for the nests that were removed but also for what was added: bacteria, faeces, litter and sound. Drawn by the desires of distant consumers, harvesters started to camp in the cave, turning an ark of endemicity into dens of gambling and drug-taking, the sound of music smothering the swiftlets' clicks. In a desperate attempt to restore order, the government sought to seal the cave with iron gates and even resorted to sending the army. 'When I went there at that time,' recalled Lord Cranbrook, Harrisson's one-time assistant in the Sarawak Museum, 'everything was stinking; everything was covered in guano. And it was full of soldiers.'

The story of Niah was replayed all over Asia: in the caves of Thailand, Vietnam, Indonesia, the Nicobar Islands and the Philippines. During the birds' nest boom, harvesting practices were exported to any site where the birds could be found, from the most remote inland caves to the most perilous sea cliffs. An inversion of Gavin Maxwell's utopian dream, this new form of intensive harvesting turned some of the most remote spaces on earth into dwellings for harvesters, drawn by the promise of the nests' exchange value. It is now hard to find any caves with large populations of swiftlets; Lubang Perintah, the cave I visited with Nuar, remains an exception, the result of an experiment between local harvesters and the state government, a small beacon of optimism.

The Icelandic author Andri Snær Magnason wrote to me that he'd considered writing a short story about an eiderdown duvet priced at $50,000. 'It'd be aimed at the ultra rich who come to Iceland, say Bill Gates or the Kardashians,' he told me. 'Then I started to think of the success – if this would actually work – how it would imbalance all the families that allow the father's sister to do the toil and get the "profit" and how this could become a microcosm for how we have messed up our resources.' He never wrote the story, but his outline spoke of the perils of a sudden rise in demand for a scarce commodity tearing up family bonds or prompting ecological collapse. In Iceland the story remains fictional, but in Niah it took on a concrete historical form.

When I read about Niah or spoke with those who had witnessed its destruction in the 1980s and 1990s, it felt as if I was back in Canary Wharf reading about extraction. Instead of discovering a new secret, say, the nature of the missing link, I had found a predictable narrative, that of the encounter between limited resources and unlimited desires. Like exhausted mines or remote islands stripped of mineral phosphates, the caves were transformed into abandoned sites of extraction. Yet I was also overcome by a sense of strangeness: the cause of this great ransacking was not the need for energy from coal, or warmth from eiderdown, but the taste of a dissolved nest. Sitting in the Great Cave, I started to fret about Iceland, wondering whether its bonds and its people's connection to the land were contingent on price rather than permanently welded in time.

THE WHITE-NEST SWIFTLET

If you leave the Penn Museum, head east along South Street, cross the Schuylkill River into central Philadelphia, cut south and walk along Washington Avenue, you find a handful of supermarkets offering South East Asian and Chinese food. My favorite is Hung Vuong, a Vietnamese supermarket, which lies between Eleventh and Twelfth. When I first arrived in the city, I'd spend hours there, browsing the different foods on offer: spiked durian the size of basketballs, dried fish, live fish, moon cakes and birds' nests, cushioned in velvet-covered boxes, presented as if they were gold medals. Compact and light, they could be discreetly slipped into an ally's hand or back pocket, satisfying a debt or creating one anew.

I picked up a box of nests in Hung Vuong. The nests in this shop were different from the black nest I had examined in Nuar's domain in Sarawak. Held in the hand, these white nests seemed a piece of bleached crust. Brittle to the touch, they could pass as prawn crackers. But when I raised them to the light, I saw that they were made of countless fine strands as translucent as gelatine. Crafted of liquid mucus, they are pure adhesive, a sculpture that can bind to the most uneven of surfaces. I thought about buying them, but then I looked at the price, $99, for three. I wondered how these nests had come to inhabit a shallow container instead of the depths of a cave, the inaccessibility of limestone replaced by the exclusivity of price.

Finding these nests in New York, Chicago or Philadelphia, I was always astounded by both their quantity and their derivatives: birds' nest drinks, sweets, skin creams and even coffee. I'd looked for the nests in remote places, entering caves in Borneo, but now the cities seemed awash with nests as if they were upmarket soap or chocolates. In the nineteenth century, John Crawfurd, a Scottish diplomat and traveller, observed that the quantity of birds' nests produced was 'by nature limited and incapable of being augmented'. Yet I noticed that there were abundant quantities of nests; their supply somehow seemed *unlimited*. The swiftlet specialist Lord Cranbrook told me that a new form of production had been invented, overcoming the natural limitations of the cave. The harvesters had started building bird hives, drawing them into towers, towns and cities. 'The birds just came', he said, 'and people opened their homes to them.'

In 2007 a group of Israeli archaeologists excavating Tel Rehov, an Iron Age settlement in the Beth Shean Valley, found about twenty-five unfired clay cylinders, each one around a metre long. Inside, archaeologists discovered the remains of honeycombs and the bees that had built them. It is hard to say when humans started keeping bees, as opposed to hunting their honey in the wild. References to beekeeping can be found in ancient Egypt; the walls of the sun temple of Ne-user-re from the Fifth Dynasty depict workers blowing smoke into hives while they remove honeycombs. But the beehives at Tel Rehov, dating to the ninth to tenth century BC, are now believed to be the oldest examples ever discovered, suggesting an annual production of up to half a metric ton of honey at the site.

This history proved useful when I read about the origins of the white nests that I'd found in Hung Vuong supermarket. Unlike the nests of Niah, they are built by the white-nest swiftlet, a different species, which relies only on its saliva to build nests. Lord Cranbrook told me that a few birds had made the first move, flying from caves into homes in Sedayu in East Java in the 1880s. It was reportedly a sign of good luck, and little attempt was made to respond to the birds until the middle of the twentieth century, when homeowners started to convert houses to suit the birds' predilection for darkness. After the price rise of the 1980s, the house-farm industry took off; Javanese businessmen started attracting birds with electronic birdsong, altering the conditions within houses to mimic caves, with quite spectacular results.

In London, great campaigns are waged against the descendants of rock doves, the city pigeons, which coat buildings and monuments with their acidic white droppings. Fed up with the damage, the Greater London Authority banned the feeding of the birds, hunted them with Harris's hawks, and placed spikes on their landing spots. Unlike British city councils, harvesters have sought to draw birds to cities in South East Asia. Spurred by high nest prices, they have converted flats, shops, hotels and banks into bird houses, altering the skyline of cities and playing bird calls over the roar of traffic. The total value of the nests produced in these houses is estimated to lie in the billions of dollars, far overshadowing production in caves.

Like caves, the surface of bird houses gives little hint as to what lies within; banks, flats and shops may all sit below a bird house, a concrete wall separating the rhythms of commerce or domestic life from a swiftlet's life cycle. Walking through cities in Malaysia, I learned to spot them, hunting for blacked-out windows, the sound of bird calls and the presence of birds, rising like smoke from roofs. But they proved almost impossible to enter. When I asked bird house owners to open up their houses to me, I received responses that were, variously, similar to what one would expect from a request to inspect their bank statement, wardrobe or internet search history. Private places, bird houses hold secrets about their owners, their wealth, tastes and capacities; the great fear of bird house owners is that a competitor will steal their intellectual property, whether the design of a particular house, the best location for a tower or the precise type of birdsong employed. Frustrated, I wrote to Lord Cranbrook for advice.

The life of the zoologist Lord Cranbrook is closely intertwined with the swiftlets to which he has dedicated himself. In 1956, as a recent graduate in natural and moral sciences, he met Tom

Harrisson at a party. Harrisson, drunk and in 'ebullient spirits', offered him a job at the Sarawak Museum. Keen to escape England, Cranbrook sailed to Singapore and then Kuching to take up his post as Harrisson's assistant in the summer of 1956. 'I just couldn't bear the thought of settling down,' he recalled. 'I had to escape.' Like other young men who had sailed to Sarawak from England, Cranbrook spoke with the accent of the British upper class and was heir to an estate, but he was different from the uniformed officials who made up Brooke's personal fiefdom. Rather than Sir Stamford Raffles, the founder of Singapore, or James Brooke, Cranbrook looked to Alfred Russel Wallace for inspiration. Like Wallace, he became intrigued by seemingly small details that could crack open the great mysteries of the natural world: the precise functions of swiftlet clicks, the range of the Bornean tiger, or the date of the domestication of the wild pig.

When he arrived in Borneo, Cranbrook moved in next to Harrisson in Kuching, but it was to Niah that he was eventually drawn. Long before the palm oil and timber boom, Niah offered him the perfect chance for adventure and escape. He lived among the Iban and Penan, sleeping anywhere and eating anything, from bats to flying foxes ('quite gamy', he recalled). Tribesmen marked him as one of their own, tattooing him with two Bornean roses on his shoulders. Sensing his ability, Harrisson set him to work on Borneo's swiftlets at the Great Cave. With little more than a rattan rope, a bird net, and a pair of plimsolls, he explored Borneo's caves, squeezing his slender frame through limestone passages to count nests and swiftlets.

Tourists are used to caves with large open entrances, but many of the caves in Borneo are entered through small sinkholes, crevices or gashes in the limestone. Covered as they often are in vegetation, it is not uncommon for livestock to fall into them, splitting their heads on the limestone below. Reading Cranbrook's notes on his trips to these caves, one gets the impression of an adventurer, willing to brave any peril to see a bird's nest. He trudged through the guano, descended into sinkholes, and yet he always brought back more than nests, an object for a cabinet or museum. He visited hundreds, if not thousands, of caves, and carefully recorded the numbers of birds and bats, gradually untangled the mysteries of their life cycles, and published countless papers. Shortly after his first visit, he became intrigued by the clicking sound and set out to establish its function. He released a few birds in the darkroom of the Sarawak Museum, turned the lights on and listened. 'The clicking stopped,' he said. 'And so it had to be echolocation!'

Whenever I spent time with Cranbrook, it was hard to get him to talk about himself. On one occasion I met him in an Italian restaurant by London's Victoria Embankment. He was down for the day from the family estate in Suffolk and wore a chalk-stripe suit, immaculately pressed, and spoke in precise, clipped tones. A stranger might have mistaken him on the street for a member of the London establishment, a retired diplomat or judge, out for a stroll before retiring to his club in St James's. I took a certain pleasure in knowing otherwise: that the large hands holding the cutlery were the same ones that had extended the wings of a bat in Harrisson's documentary; that beneath the chalk stripes lay two Bornean roses, tattooed by the Iban; that he cared more for the precise means of swiftlet echolocation than politics, law or the workings of power.

In the 1980s and 1990s, Cranbrook began to spend more and more time in cities, tracing the passage of the birds into houses in South East Asia. He met the new men and women of the industry, many of whom had grown rich off the trade in swiftlet saliva. One

of his contacts in the world of birds' nests was an Indonesian businessman named Dr Boedi Mranata. A biologist by training, Dr
Boedi was among the first to pioneer early techniques in swiftlet
farming in the 1990s, experimenting with different house designs
and bird calls. Drawing on his knowledge of swiftlet biology, he
proved fantastically successful, attracting huge numbers of birds to
his houses. Now one of Indonesia's wealthy men, he controls an
empire of swiftlet hotels stretching from Java to Kalimantan, and supplies nests to China and Hong Kong under his brand, Xiao Niao, or
Pristine Nests. The website of Xiao Niao describes Dr Boedi as the
King of Birds' Nests, a title whose aptness I would soon appreciate.

One summer I joined Lord Cranbrook on a trip to Jakarta to visit
Dr Boedi, to take samples from his house as part of the British zoologist's research into the origins of house swiftlets. The day after we
arrived, Dr Boedi arranged for us to be picked up from our hotel and
chauffeured back to his home, a large building in a gated community
on the city's outskirts. Stuffed full of treasures, his home resembled a
miniature museum of unabashed opulence, at once a celebration of
his intellectual pursuits and a monument to his success in the birds'
nest industry. We wandered through it like schoolchildren, admiring
old cannons, stroking a fly whisk made from a yak's white tail, running our hands along hardwood furniture, staring at a vast collection
of Chinese ceramics in a sealed vault and a reconstructed wooden
temple outside. Nearby were an enormous fossilised tree trunk and a
peacock, whose call rose over the trickle of a fountain in a tasteful,
simple rock garden.

A short, soft-spoken man with glasses, Dr Boedi has the air
of an intellectual, an academic biologist. Like Lord Cranbrook, he
seemed drawn to knowledge for its own sake, enthusing at length
about the life cycle of swiftlets and the exploits of early Chinese

merchants. A renowned expert on ceramics, he'd recently completed a detailed study on martabans, the large storage jars that the Chinese traded with Borneo's inland peoples in exchange for swiftlet nests. But behind the bookish exterior I occasionally caught glimpses of a different character, the robust operator who has remained at the top of a notoriously ruthless industry for the best part of three decades.

Shortly after we arrived, Lord Cranbrook gave a talk on the domestication of swiftlets in Dr Boedi's living room. The room was packed with Indonesian students, biologists and businesspeople, all seated in a circle around the zoologist. In addition to Dr Boedi there were two other Indonesian birds' nest magnates, who together with him had divided up much of Indonesia among themselves, forming the 'Big Three' of birds' nest harvesting. Like him, they possessed great self-confidence, but they deferred to him as if he were a wise teacher or guru. When Dr Boedi cracked a joke, as he often did, I noted that everyone in his entourage laughed a little too loudly, with the exception of Lord Cranbrook, who merely grinned. 'Why does not Lord Cranbrook laugh?' asked one of the Big Three, clearly perplexed. 'We always try to make him laugh.'

These birds' nest businessmen seemed so different from Nuar, the climber at Niah. Most comfortable in batik shirts and Italian leather shoes, they specialised in the vagaries of the Chinese luxury market, fluctuations in currency exchange rates, pricing models and security systems for birds' nest houses. Seated at their desks in Jakarta or Medan, they could monitor their bird houses from their offices, rarely lifting their own body weight, let alone scaling a *belian* pole into a limestone dome. They were strangers to the cave, and the wild birds it held within, untrusting of limestone's darkness and smells. I saw this when we later all visited a cave in Java to take further samples over an afternoon. By the time we left the cave, Cranbrook's T-shirt was drenched with sweat and stained with dust, but their batik shirts were unsullied, the Italian leather of their shoes unscuffed.

The mood in Dr Boedi's living room was buoyant. A few months back, China had lifted a ban on birds' nests from Indonesia, in place since 2011, after health concerns ostensibly relating to the levels of nitrite in the nests. As the head of the Indonesian Birds' Nest Traders and Farmers Association, Dr Boedi had been at the forefront of efforts to get the ban lifted, lobbying Chinese authorities. Now that he had succeeded, Indonesian suppliers were making preparations to sell directly to the Chinese, reaping the rewards of an unmediated supply chain. Over the chatter in the living room, I could hear Dr Boedi's two children, Harry and Ariani, busily discussing marketing strategies, including commissioning videos about their father's nests for the Chinese market.

Later that week we travelled to Dr Boedi's bird house in western Java. After a journey of a couple of hours, we reached a plot of land full of coconut and guava trees, a rare island of green away from the concrete and tarmac of Java. I never learned its precise location, though it clearly lay close to a zone of intense volcanic activity, as testified by the smell of sulphur from bubbling springs. At the heart of the plot lay Dr Boedi's creation. A brutalist block, it resembled a huge pillbox or bunker shooting out tracer fire of swiftlets from an opening at its top. Around its base there was a small channel of water, a defensive moat, designed to deter insects or rodents. 'I'm not really an early bird,' Ariani told me, 'but when I [come to these houses] I would love to wake up [early] because at sunrise you would see all these birds swarming out. It's like apocalyptic.'

Dr Boedi stood before it with his arms outstretched. His mood had softened, his deadpan expression replaced by a joy that was almost child-like. After stepping over the miniature moat, he opened the metal door to the tower and invited us into his lair. We were greeted by warm moist air, swiftlet clicks and the sound of birdsong, played from sets of speakers suspended above us. Gummed against wooden beams were hundreds of nests, stuffed full of baby birds, dark black against the whiteness of swiftlet saliva. Dr Boedi beckoned us over to one of the beams and placed his finger inside a white nest, where it was mouthed by a swiftlet chick. He began to whisper gently to the birds. 'Let's say', Dr Boedi said, 'some houses can harvest only two hundred kilos – sometimes four hundred kilos [per two months]. One kilo is one thousand dollars. So two hundred thousand dollars every sixty days. It's a small bank!'

When apiarists first set out to harvest honey, for some thirty dollars they can order a batch of queens online that can be installed in a hive; it is normally a straightforward and relatively inexpensive procedure for experienced keepers. By contrast, the birds' nest harvest presents a different world; although some harvesters may place fertilised swiftlet eggs in houses, swiftlets must be seduced, drawn to houses by birdcalls or the prospect of attractive lodging. Before building a house, harvesters can first test a site by playing a recording of swiftlet chicks in distress. If the site is promising and swiftlets appear, a tower is built, and speakers are installed that blast bird calls to attract residents. There is a thriving market for swiftlet sound, with outlets offering recordings of chicks, mating birds or distress calls with curious names like 'Black Cloud' or 'Baby King'. Salesmen even offer items that allegedly attract swiftlets, from plastic nests to pheromone sprays.

In the nineteenth century there was a great proliferation of theories about beekeeping techniques. Beekeepers would experiment with different hive designs, the best ways to draw the insects. Similarly, Dr Boedi gave great thought to the conditions in his bird houses, monitoring the levels of humidity, light and sound. Ariani said he had some sort of 'Dr Dolittle' knack, based largely on instinct. 'I've gone to birds houses since I was a kid and I would still go there and not understand whether it is too dry or too humid. You know, my dad would like [say], "Just feel it on your neck." And I'm like, "What do you mean, from your neck?" "The sweat."'

An apiarist may also create varied flavours of honey, releasing bees in fields of blossom: lavender, alfalfa, blueberry or eucalyptus. Like beekeepers, harvesters can create different types of nest in their houses by tinkering with their conditions. 'We're trying to produce orange nests because that's the demand from Singapore right now,' Ariani told me. 'It's what we call the C4 colour.' Red or orange nests can be five times the value of white nests, and the practice of creating them is shrouded in secrecy. Some believe it is the birds' blood that colors the nest, but it is in fact the result of nitrifying bacteria within the nests reacting with ammonia from guano that is present in houses. Ariani, a good businesswoman, would not reveal any specifics: 'It has a lot to do with the condition of the house, how humid it is, the temperature, a lot of chemical reaction.'

We proceeded upwards into the tower's second level, climbing a set of guano-splatted stairs up to a floor with a small pool of filthy water. The birds, I noted, shot through each floor via great manholes, heading toward the beams of light at the top of the tower. I carried on counting the nests, but soon paused, my attention directed towards the imperative of not slipping or falling. What was this place? A farm? A bank? A hive? The word *hive* is related to the Old Norse *hufr*, or hull of a ship. Open vessels, Dr Boedi's houses could lose their inhabitants, destroying the fortunes of their build-ers. 'You cannot put the birds' house in the bank as your asset [col-lateral],' Dr Boedi explained. 'Tomorrow they might leave.' I thought back to his large collection of antiquities, locked in his secure room with an alarm; unlike swiftlets, Ming ceramics are not prone to cap-ital flight.

I reflected on the nature of the harvest at Niah, how different it was from this operation. Within the caves the birds did not need Nuar. Indifferent to him, they tolerated his presence as long as he did not cook onions, play loud music or take too many nests. Here I saw that the harvest was no longer a one-sided bargain: much like the eiders, the swiftlets now needed the harvesters, dependent upon them for accommodation and sometimes protection (some nest har-vesters will kill predators, such as owls). But the swiftlets still re-tained some control, capable of leaving their houses if their landlords breached their contract. 'It's not really about how to milk the most money out of this animal,' Ariani told me. 'You get all the birds nest-ing and popping eggs and they are going to find their eggs don't exist anymore and their house got robbed. So they are probably going to run away.' Instead, new houses are left alone, until they have five thousand nests. 'When it reaches ten thousand [nests], you're safe to harvest every sixty days,' she said. 'You want swarming houses.'

After our tour of his tower, Dr Boedi laid out refreshments un-der a gazebo made of coconut palm fronds. While we feasted on coconuts and guavas in the shade, Ariani and her brother Harry detailed plans for their family business. They showed me a market-ing video on their laptop, featuring their bird towers in a remote part of Kalimantan, Borneo, and spoke of their hopes that birds' nests would make it big in Europe and America. 'When there is a proper paper that comes out that says [there are health benefits], I'm sure it is going to be commercialised,' Ariani enthused. 'It's going to be turned into pills, canned food and jams.'

The question of the nests' medicinal properties is both highly emotive and commercially sensitive. An ever-growing body of research in China and Japan indicates the possible health benefits of birds' nests. Papers point out that the nests contain sialic acid, which is present in human milk, and that the nests may have immuno-enhancing effects, or an epidermal growth factor-like (EGF) hormone. So far, however, to my knowledge, no peer-reviewed journal in Europe or America has attributed any medicinal properties to the nests. Lord Cranbrook, for his part, tactfully skirted the subject while we visited Dr Boedi's house. 'Well, eating a nest certainly won't do you any *harm*,' he said quietly.

I didn't mention any of this to Ariani, nor did I feel it my place to do so. I felt grateful for her presence, her quiet explanations of the vagaries of this business. After school in Jakarta, she went to business school in Boston, learning to field questions from perplexed contemporaries about her father's line of work. 'They would imagine birds' nests, you know, like all made out of twigs in their backyard and, you know, how does that make money?' Like those students from Boston, I felt as though she was taking me under her wing, empathising with my confusion. At times even she was befuddled by the logic of birds' nest harvesting. 'I wish my dad would just take over a restaurant,' she confided. 'It'd be so much easier for me to take over.'

By contrast, Dr Boedi was tight-lipped, keeping his cards close to his vest. When I once asked him a specific question outside a cave, he replied by giving me a shard of ceramic he found on the ground. 'Probably Ming,' he mused. Even with the kind assistance of Ariani, I left Jakarta feeling that I had learned little about how

Dr Boedi had built his fortune. It was only later, long after I left Java, that I wondered whether we had not been gently directed by the King of Birds' Nests. Never unaccompanied, never left to pick up a bill, never unobserved, we had fallen under his spell, dancing to his tune like the thousands of swiftlets drawn to his towers.

In the nineteenth century, Lorenzo Langstroth, an American clergyman, developed what would become the standard beehive. It had removable frames, which allowed beekeepers to inspect the bees and harvest honey without harming them. 'The chief peculiarity in my hives . . .', he wrote, 'was the facility with which these bars could be removed without enraging the bees.' Dr Boedi, it seemed, had devised a system that embodied Langstroth's spirit, identifying his own interest with that of the birds. Wary of losing the swiftlets, he removed the nests only once the birds no longer had any use for them.

If the harvesters' relationship with these birds changes, then it is unlikely to be because of individual indiscretion, but is instead a result of collective malaise. The Indonesian island of Java was once the heart of the swiftlet industry, the home of its pioneers and vast houses. But with the island's rising population, devastating forest fires and urban expansion, the birds gradually deserted its city houses, crossing the South China Sea to towns in Borneo. Even their tenancy on this great island is not assured. Borneo is not subject to the same rapid population growth as Java, but the intensive cultivation of palms for oil relies on pesticides that kill the birds' food source. If palm oil continues to expand, replacing primary forest with uniform rows of crops, Borneo may just be another stepping-stone for these birds on their way to the last untouched areas of South East Asia.

As I travelled through Indonesia and Malaysia, harvesters told me that the birds were vacating properties in large numbers, leaving them 'dead'. 'My dad is convinced that it's because of the city life that it's pressing its way to the more pristine areas,' Ariani told me. 'Now the air quality has changed and the population [of birds] has drastically decreased.' In the city of Johor Bahru, on the southern end of Peninsular Malaysia, I saw rows of deserted bird houses, squalid flats surrounded by sewage and rusting cars, their tweeters blasting birdsong into the empty sky. Once valuable investments, they were now unfit for humans or birds. 'It's the talk of the town,' said Lim Chan Koon, who now consults on bird houses. In bars, birds' nest associations, university departments, company offices and online forums, the debate raged as to why they were leaving. 'We have no data, just observation.'

As the birdhouses emptied, small-time harvesters gave up, turning to other trades to serve the Chinese market. Dr Lim started farming empurau, a large freshwater fish that is among Malaysia's most expensive. Outside Kuching, Sarawak, he showed me his capacious tanks, where the fish circled in crystal-clear water. He'd had enough of the unpredictability and risk of the swiftlet business, but others, like Dr Boedi, had decided to chase the birds to wilder areas in the pursuit of greater profits. In Kalimantan, Borneo, he had a team of scouts tracking the birds and reporting on their movements. Dr Boedi bought up as much land as he could, building huge towers in the swiftlets' pathways. Once he built a house, Ariani told me, other harvesters would take note and seek to intercept his birds. 'All these birds' nest farmers, wherever my dad builds a house, they follow. So you would see, we'd build a house and then within two years . . . there are like five attacking us.' I imagined Dr Boedi following the birds until the whole of East Asia became oil palm and cities, dancing this strange waltz, sometimes leading, sometimes following.

AERODRAMUS INEXPECTATUS GERMANI—
THE GREY-RUMPED SWIFTLET

In his essay 'Some Thoughts on the Common Toad', George Orwell enthuses about the delights of nature in urban spaces, the varied bird life to be found in London: 'I have seen a kestrel flying over the Deptford gasworks, and I have heard a first-rate performance by a blackbird in the Euston Road. There must be some hundreds of thousands, if not millions, of birds living inside the four-mile radius, and it is rather a pleasing thought that none of them pays a halfpenny of rent.' No such logic exists in many cities in Malaysia, Thailand, Indonesia and Vietnam; there, it is the swiftlets that dominate the skyline, paying their rent with their valuable nests.

I'd watch them funnel into converted flats, swarming as they had done in Nuar's domain. Like sparrows, the swiftlets now seemed part of the city, their troglodytic history erased. One day, perhaps, their clicks will become redundant, vestigial reminders of the caves' intricate contours. I thought of the pigeons that populate cities around the world, feeding on refuse and nesting in buildings. It is easy to forget that their ancestors, rock doves, once nested high up on cliffs, surveying the sea below.

Cranbrook observed these city swiftlets with great interest, tracing their movements throughout the region. During hundreds, if not thousands, of visits to bird houses, he noted that house birds were very different from swiftlets nesting in caves; they had different coloration and had no interest in returning to limestone. It seemed possible, he thought, that a new species had formed, one that was genetically distinct from the cave-dwelling birds and that was hard-wired to choose concrete over limestone. 'What we are seeing is the most recent form of domestication,' he told me. It was strange to think of swiftlets in the company of dogs, pigs or bees, creatures domesticated for comfort, flesh or sweetness rather than the desire for saliva.

In recent decades, zoologists, botanists, archaeologists and ge-neticists have sought to find the wild ancestors of domesticated an-imals and plants, from cows to cats to corn. By comparing samples of ancient DNA with those of living animals and plants, they have shed light on questions once thought to lie beyond our understand-ing. Thanks to their research, we now know, for example, that all taurine cattle may be descended from a single herd of eighty ani-mals that lived some 10,500 years ago in Mesopotamia, or that Przewalski's horse, once considered to be the last wild species of horse, is in fact the feral descendant of horses domesticated by the Botai culture, a group of hunters and herders that lived in present-day Kazakhstan some five thousand years ago. Visiting bird houses, Cranbrook was curious to know the origin of these house-farmed swiftlets. Where did they come from?

The hunts for wild ancestors serve more than a taxonomic importance, the desire to fill in a family tree or use the correct no-menclature. As agriculture turns to monoculture, we have lost much of the genetic variation in our plants and animals, making them vulnerable to collapse, as proved by the Irish potato famine of the 1840s or the banana blight in the mid-twentieth century. Cran-brook was concerned that house-farmed swiftlets, like these domes-ticated species, were particularly vulnerable. Preliminary studies of

Malaysian and Vietnamese birds in house farms showed that there was very little genetic variation among them; he feared that this could lead to outbreaks in disease, devastating birds and producers alike. Like escaped farmed salmon or shrimp, these house-farmed birds might breed with wild populations of birds, further complicating the genetic picture.

One of the putative ancestors Cranbrook identified was a beautiful small swiftlet, *Aerodramus inexpectatus germani*, or grey-rumped swiftlet. He called the bird an 'island super-tramp' because of its ability to rove between islands, from Hainan to Vietnam, Malaysia to the Andaman Islands. Unlike the black-nest swiftlets of Niah, the grey-rump is one of two swiftlet species that produces the highly prized white nests. While Cranbrook was a student, the birds were abundant, but after the price increases of the 1980s, many of Borneo's sea caves, stacks and cliffs were ransacked, denuded of the nests by landowners or thieves. Cranbrook now believes that they were among Borneo's most threatened birds. When he spoke about the grey-rump, he often used the word 'tiny', uttered with such delicateness as to suggest fragility and rarity.

In October 2014 he had sent a couple of nimble students to explore the coastal caves of Sabah, the Malaysian state next to Sarawak, and check on their bird populations. In all the sites they visited, the students found only two caves with grey-rumped swiftlets, on two small islands off Borneo's north-western coast, Mantanani Besar and Balambangan. Now, as part of a British Malaysian study on swiftlet origins, he planned to visit the caves, catch a grey-rump, and take a sample feather, with the assistance of a small party including the geneticist Dr Sarah Ball and the filmmaker Jamie Curtis Hayward. When Cranbrook casually invited me along, I was slightly reluctant, given that he had described the cave in question as 'difficult, but not impossible to access', but I took some consolation from the fact that he was shortly due to turn eighty-two, an age when presumably he was past his climbing prime.

Mantanani is a small island, no more than three kilometres long, rising out of turquoise water. It is inhabited by a community of Ubian Muslims, their ranks vastly outnumbered by the hundreds of low-budget tourists from China, drawn to the island's bright white beaches, scuba diving and karaoke bars. We arrived there in the morning, having taken a boat from Sabah's coast. Bearing bird nets, we walked among the tourists on the beach and past pink sunbathers, carpets of drying fish, monitor lizards and hundreds upon hundreds of plastic bottles brought in by the tides, whose frequency Cranbrook began to calculate per square metre.

Cranbrook's target cave lay on the north-western tip of the island at the top of a limestone hill, far from the bustle of Chinese tourists and scuba diving tours. Known as the Governor's Cave, it was the only one of the island's sixteen caves that still had a population of swiftlets. Like most assets with a valuable yield, the cave was owned by a distant landlord but managed locally by a guardian. His name was Mohd Salleh. A solitary figure, Salleh lived in a small cabin removed from the island's main settlement, subsisting apparently on a diet of coconuts, fizzy drinks and bean gruel. A brilliant climber, he ascended coconut palms and the limestone caves with ease to harvest nuts or nests, modelling himself on Tarzan. 'He believes in Tarzan as one might believe in the Holy Ghost,' Cranbrook commented.

Whether descending into caves, climbing coconut palms or attending a local wedding, Salleh always wore a bandana and a pair of safety goggles, giving him, Cranbrook commented, the appearance of a World War One pilot. Conscious of Cranbrook's status, the guardian referred to him as Tan Sri, a Malaysian honorific roughly equivalent to a knighthood. Yet between the two men there was an informality bordering on affection. Salleh rolled with laughter at the shoes of this visiting dignitary, a pair of tattered plimsolls, which he said were favoured by timber workers. Cranbrook, in turn, teased Salleh over his choice of cinematic hero, Tarzan, explaining the harsh botanical realities of swinging from hanging vegetation. 'The problem with pulling on vines,' he said, 'is that it releases a lot of dead material.'

After much discussion with Cranbrook, Salleh offered to lead us to the cave. We set off towards the limestone hill through an abandoned coconut plantation, our feet rolling on buried nuts. Aided by two walking poles, Cranbrook moved quickly in his plimsolls, though every so often he would stop, examine a leaf or a seed, and say the Latin name of the tree or plant to which it belonged. At other times his gaze was drawn down towards the thick network of roots that covered the jungle floor. 'The soil is so poor here', he explained, 'that trees desperately spread their roots over the surface to get nutrients.' Before long I found myself trailing behind him, weighed down by my heavy leather boots, while he powered onward. 'Aren't you hot in those boots?' he asked unhelpfully. I wanted protection from snakes, I said. 'There are about three snakes on this island,' he quipped, 'none of which are poisonous.'

It was a tiring climb up the side of the limestone hill, thick with primary vegetation and insects. By the time we reached the top our shirts were drenched with sweat. Our eyes stung with insect repellant and I noted that my plastic pencil, melted by DEET, now bore my fingerprints. Cranbrook suggested that Jamie and I go on ahead to assess the descent to the cave mouth, a gash in the limestone cliff some thirty metres above the sea. The night before, a storm had thrashed the limestone, which, already loosened by the repeated action of countless roots, was now slippery to the touch. It was a hard climb, and we both agreed that it was not a good idea for Cranbrook, a man nearly three times our age, to attempt it. We said as much, but Cranbrook had other ideas. 'It's called the Governor's Cave,' he said, a hint of irritation creeping into his voice, 'so presumably the governor visited it at some stage.' If a colonial governor had reached the cave in the nineteenth century, then there was no way that Cranbrook would not.

In Victorian times it was common for explorers to die or be maimed when trying to collect specimens in the tropics. Jamie and I looked at Cranbrook, his grey shirt now almost black with sweat and his red handkerchief resting on his head. I did wonder whether we might be the latest additions to this list, and yet the zoologist's good judgement was evidenced by the simple fact that he was still alive after half a century of exploring the caves. 'I have planted trees that have died before me,' he once joked.

The thickness of the stem of a banana plant is no indication of its sturdiness; the Malaysian poison tree cannot easily be distinguished from a myriad of others; and the roots of the pandanus tree have spikes. All these botanical facts I learned during my descent toward the cave's mouth, grasping for something to hold on to. Above me Cranbrook's plimsolls kicked around for footholds, his weight partially supported by Salleh's antique rattan rope. Below loomed the cave opening itself, a disincentive to slip, if one was needed, as a tumble would take you into its guano-lined belly.

Sweating and exhausted, we huddled on the rock ledge at the cave mouth. Cranbrook tied a bird net to two vines that Salleh had hacked, and propped it up close to the cave entrance. It was difficult, fiddly work, complicated by extreme humidity and swarms of insects. Until this point, I had served little use on this trip, but now came the moment to use my single asset, a comparatively youthful pair of legs. At Cranbrook's request, Jamie and I descended into the cave to count the nests. Scrambling down the narrow passage armed with bird nets, I heard Salleh's voice, followed by his piercing laughter. 'Snakey, snakey!' Pythons are often found in these caves, feeding off the bats and birds, but it was unclear whether the guardian was warning us of their presence or having a joke at our expense.

Unlike in Niah, little light entered this cave, its narrow confines lit only by the beams of our torches. Devoid of lichen, the cave was pale, almost corpse-like. The only indication that it was not dead, empty of life, was the occasional *click click* of a swiftlet navigating in the darkness. Scanning the cave walls meticulously, Jamie was distinctly calmer in the claustrophobic space of limestone. Unlike me, he had experience working as a field zoologist in Kalimantan, skinning birds under Cranbrook's guidance. Unbothered by imagined risks, he soon found a small cluster of white nests and the outlines of swiftlets at the top of the cave, his attention drawn by the presence of fresh deposits of guano on the cave floor.

As in the Great Cave, I felt uncomfortable in this space, but the root of discomfort was not related to darkness or the reek of guano. Looking upwards I counted only two small cups in a small cave on a small island off a large island. It felt as if we had opened up a series of Russian dolls only to reach the inmost, smallest vessel and discover that it had already been cracked open, its contents emptied and shipped, most probably, to the Chinese mainland or Hong Kong. I began to question, too, what we were doing here, worrying that by measuring or representing these birds we might end up finishing them off.

Such anxieties left me when I saw Cranbrook handle the minute creatures. By the time we emerged from the cave, he was carefully removing a bird, which fluttered like a fish, from the net. In his large hands, it appeared insectile in scale, its black eyes dominating its head. Grinning, Cranbrook placed the bird inside a cream cotton bag bearing the name of an upmarket Swiss shoemaker. Next to the bag I saw two other shoe bags gently pulsating in the shade; the fishing trip had been a success. Lacking much space at the cave mouth, Cranbrook opted to proceed to the top of the cliff to take samples. Drenched with sweat, we scrambled up the limestone, the shoe bags attached to Jamie's belt like gunpowder pouches. At the top Cranbrook opened the bags, took measurements of the birds, removed one small feather from each, and passed the feathers to Sarah, who placed them in sample bottles. 'You're a bright fellow,' Cranbrook whispered, before releasing one bird.

Cranbrook is not an emotional man, but I detected a hint of sadness in his voice as he removed the small feathers from these miniature birds. In the late nineteenth century, birds' nests had been among the great commodities of Sabah, 19 per cent of the export revenue of the North Borneo Chartered Company, but now it was clear that most of the caves off this great island were unpainted with swiftlet saliva. Even as house-farmed birds have proliferated, spreading all over Asia, the numbers of cave swiftlets have not recovered. I thought of the wild salmon, whose numbers continue to plummet as farmed stocks in sea cages increase exponentially. With every step that I took from the cave, the birds' nest trade became more and more familiar, almost morphing into any other market, from caviar to salmon to cattle. Familiar, that is, until I reached the final stage, the eating of the nest itself.

On my way back home, I stopped off in Johor Bahru to meet Dr Tan Boon Siong, a birds' nest processor, who works out of a business centre that he shares with a cosmetics company. A long-time devotee of birds' nests, he traces his interest in them back to the time when his mother was seriously ill with cancer. While she lay in the hospital, Dr Tan fed her bird's nest soup, having heard of its extensive curative properties. His mother, he said, got better, and since then he drank it every night, a sort of anti-ageing cocoa. 'That is why I am sixty years old and still look young,' he said.

He led me into his basement, a spotless space as clean as the collections viewing room at the Penn Museum. It was here that he processed the nests, most of which would head across the straits to Johor Bahru's neighbour, Singapore, where they would be eaten by wealthy Chinese. Keen that I should experience their taste first hand, Dr Tan took a nest out of its plastic box and dropped it into boiling water. Surrounded by small bubbles, the nest softened, its threads swelling to form a translucent gelatin. After half an hour it had fully dissolved, a bird's work undone. He poured it into two mugs and we drank the broth. 'It's like milk,' he said. I did not have the heart to say that I did not taste a thing.

CIVET COFFEE

n 1692, Daniel Defoe faced financial ruin. Long before he wrote *Robinson Crusoe*, he was a merchant and speculator, a 'projector' or wildcatter. The outbreak of war with France had sunk his commercial ventures as French privateers attacked British trading ships, inflation rose, and Britain's merchants felt the strain. Pursued by no fewer than 140 creditors, he owed £17,000 after a series of business failures. In Defoe's time there was no possibility of discharge from bankruptcy, and little remained between him and the debtors' prison. Harried by the threat of imprisonment, he resorted to a series of improbable business schemes. In the course of his mercantile career, Defoe sold wines and liquors, and men's undergarments; imported firewood, tobacco and snuff; owned a factory that made roofing tiles; and invested in a diving bell to find sunken treasure. But nothing could rival the strangeness of his acquisition, for £852, of some seventy civets, small nocturnal creatures found in Africa and South Asia.

Notoriously difficult to classify, the civet has been compared to at least five other mammals. In *The Historie of Foure-Footed Beastes*, published in 1607, Edward Topsell quotes the following description:

It is greater than any Cat, and lesser then a Taxus [badger], having a sharp face like a Martin, a short, round, blunt

eare, which was black without but pale within, and on the
brims, a blew skye-coloured eye, a foote and Legge blacke,
and more broad or open then a Cats: Likewise a blacke
claw, neither so crooked nor so hid in the foote as it is
in a Cat, but their teeth are more fearefull and horrible.

Since Topsell's description of the civet, debate continued to rage
about its proper place in the animal kingdom. In 1752 the British
physician John Hill concluded that it 'was first supposed of the cat
[genus], and afterwards of the dog kind, but it is truly one of the
badger species.' By 1821, John Edward Gray had settled the matter,
placing civets in their own separate group, the viverrids, which today
includes some thirty-eight species.

That Defoe should invest in this unclassifiable creature might
seem strange, but it would not have surprised his contemporaries in
seventeenth-century England. One of the civet's identifying features is a 'pouch or gland' below its anus, which secretes a yellowish
liquid the consistency of honey or butter. Its smell, variously compared to that of cheese, earth or bodily fluids, is highly potent and
had long been sought by perfumers as a base or fixative for scents. It
was typically diluted and smeared on wigs, sheets or snuff, and its
smell evoked sophistication and opulence. 'There has been a time
when the produce of the civet's posteriors was in the highest estimation with the ladies and with effeminate men,' wrote William Fordyce
Mavor, a teacher turned clergyman, in 1800.

Much like the bird's nest, the liquid transcended its physical limitations to become an elixir. An antidote to the smells, diseases and
strains of city life, it was believed to be a cure for anything from de-
pression to erectile dysfunction. 'Give me an ounce of civet, good

apothecary, to sweeten my imagination,' cries Shakespeare's King Lear in a fit of lust. In the sixteenth century, the Sienese writer Pier Andrea Mattioli proposed that civet be rubbed on the navel to treat 'strangulation of the uterus' or be smeared on the penis to increase the pleasure of intercourse.

By Georgian times, civet had become one of the most expensive consumer goods in Britain, more prized than bear grease or turtle. According to the historian Christopher Plumb, a grain of civet, sufficient to scent linen or a day's worth of eau de toilette, sold for two pence, the same price as a day's supply of coal or a hot meal in London; a single civet, the creature, was estimated to yield £2 worth a year of the liquid. Typically it was imported into Britain from Cairo, Basra, Calcutta and Ethiopia, but liquid harvested in the Netherlands or within the British Isles fetched a higher price, its origin a reassurance against adulteration. No surprise, then, that Defoe should have offered about £12 for each civet.

What a curious thought, to imagine this creature playing an unseen part in the moves and intricacies of seduction and social manoeuvrex in dance halls, country houses or royal courts. Like beavers or whales, whose oil or ambergris were also prized as fixatives, the civet stalked the social lives of the bourgeoisie and the aristocracy. There was something delightfully unmasking about the idea of this rich liquid, itself part of the sex lives of wild creatures, appearing in our own courtship displays, masked by politesse.

Defoe kept his civets in a house in Newington Green, now in north London, which he knew from his boyhood. According to a complaint later submitted to London's Court of Chancery, the house was 'fitted & made convenient and proper only for the keeping & feeding' of the civets, which were held in 'severall Coopes with

Troughs and Cisterns to feed them in and stoves for keeping of ffires in the severall roomes for the preservacõn of the said Catts.' Such conditions were typical for the keeping of civets: high levels of heat apparently increased their production of the secretions, inducing them to 'sweat'. The civets were confined in narrow hutches, their perineal glands scraped with a long spatula akin to a spoon used to remove bone marrow.

The harvest of birds' nests was governed by a constitution, a certain accord that prevented excessive greed or cruelty. 'The birds', Ariani Mranata had told me, 'are our boss.' Unlike the hive, civet cages offered no limitations, no natural checks on rapaciousness. Unscrupulous harvesters would often scrape too much civet from the creatures, causing great suffering or milking them to death. Pietro Castelli, a seventeenth-century Roman physician, recorded the practice of harvesting civet in Córdoba: 'One [servant] drew the chain, wherein the Zibet [civet] was tied, another held the hind legs, a third chafed the bagges, and with a large ear-pick fetched the Civet clean out, scraping on all sides, then wiped the short-hair of both bags with cotton wool.' This process was carried out three hundred times to fill a chestnut shell with liquid.

Inspect your bathroom or smell your lover, and you are unlikely to find traces of a civet. The Georgians eventually tired of its smell, disgusted by its origins or associations with excess and luxury. 'And all your courtly civet cats can vent, / Perfume to you, to me is excrement,' wrote Alexander Pope in his Satires of 1738. Instead of scraping the creatures' glands, the English turned to floral scents, which were much more acceptable to sensitive Londoners increasingly perturbed by civets' associations with excrement, decadence and sexuality. 'To the credit of taste and elegance,' noted William Fordyce Mavor in 1800, the 'very considerable' traffic in this perfume was on the decline.

Although some twentieth-century perfumes, most notably Chanel No. 5, relied on civet, chemical fixatives now dominate the market.

I felt relief at the passing of civet, the release of this creature from our social lives, its trade and captivity confined to the past. The history of materials of the twentieth century is largely one of substitution, replacement of natural materials with synthetics: of ivory by plastic, of dung by chemical fertiliser. The disappearance of civet seemed to me to be part of this process, a sign of progress in our relationship with the wild until, that is, I looked in my own kitchen. In my cupboard, nestled among old vitamin tablets, cod-liver oil and saffron, I discovered traces of a civet. Not in a glass perfume bottle, but in a plastic packet. Opening it, I found ground coffee with a fruity smell that, I read, had apparently been eaten and excreted by civets, absorbing their digestive juices. Unlike the buttery secretion sought by Defoe, this substance did not once belong to the civet, but had passed through it, tracing its innards.

In the nineteenth century, the Dutch conceived a ruthless and innovative system to extract value from their imperial possessions, which they had recently acquired from the Dutch East India Company. Under the Cultuurstelsel, or Cultivation System, introduced in 1830, farmers in Java were forced to pay revenue to the Netherlands treasury in the form of labour or agricultural produce, such as sugar, indigo or coffee. According to some coffee producers, during this period of the Cultuurstelsel local farmers first started collecting coffee beans that had been eaten and excreted by civets. The story goes that Dutch colonists forbade locals from drinking coffee made from standard beans – a valuable export commodity – and so farmers instead used civet scat to make what became known as *kopi luwak* (civet coffee). The civet 'selects only the ripest and most perfect fruits', wrote Thomas Horsfield, an American naturalist, 'and the seeds are eagerly collected by the natives, as coffee is thus obtained without the tedious process of removing its membranaceous arillus.'

As part of the processing of non-civet coffee, the cherries are typically left out to dry in the sun (dry processing) or placed in special flotation tanks (wet processing), to remove the skin and pulp that surround the bean. When a civet digests a coffee cherry, it essentially carries out the same task with its digestive juices, saving time and, as Horsfield imagined, labour. Whatever the truth of its origin stories, civet coffee – or some civet coffee – certainly has its own distinct flavour, supposedly the result of the beans' interactions with the creature's digestive enzymes. Enthusiasts variously describe it as 'musty' or 'earthy' with 'a top note of rich, dark chocolate'. Others are not so complimentary. The coffee historian Jonathan Morris notes that civet coffee is often said to have a low acidity and a lack of bitterness, descriptions that one would use to defend a bad coffee rather than praise a delicacy.

It was disconcerting to find this coffee in my own possession, the sudden reappearance of this oft-mistreated animal in my own domestic space. Inspect one's cupboard and it might resemble a cabinet of curiosities, revealing a network of different relationships. Perusing my own, I found connections to wild cod, crocus and palm oil, but I had not expected to come across such a strange object lurking in my kitchen. I thought of the Chinese diners who feast on birds' nest soup in Harrisson's documentary, or on larvae or scorpions, evoking Prince Philip's quip: 'If it has four legs and is not a chair, has wings and is not an aeroplane, or swims and is not a submarine, the Cantonese will eat it.' Reflecting on the contents of my own cupboard, I was perhaps no different, willing to eat anything.

'On paper it made sense,' Matthew Ross, a British civet coffee producer, told me enthusiastically. 'I loved the idea of the story. You get a civet cat, which has an ability to choose coffee in excess of any human. It will only eat the perfectly ripe fruit that it wants to at that time. I like to think of [the civet] as the master champagne maker walking through the vineyards, collecting the very best grapes.' Matthew dated his interest in civet coffee to his time working in Hong Kong as a derivatives broker. While taking clients out for dinner on a business trip in Japan, he was surprised by the poor quality of coffee there. 'It just became clear that there wasn't a delicacy version of coffee,' he recalled. 'That night I decided I would make the best coffee in the world.'

He returned to Hong Kong and started researching different coffees. 'I wanted scarcity, I wanted flavour, artisan, craftsmanship. I wanted it all. And then I found *kopi luwak*.' Over the next couple of years he devoted more and more time to learning about it. 'What had initially turned out as a side project just took over my life. And in 2011 I was between jobs . . . and I decided that I had to follow my heart and follow the coffee.' Others in the mainstream coffee market were not so convinced; Matthew recalls the head of a coffee association referring to *kopi luwak* as coffee that went 'from arseholes to arseholes'. But he ignored them, putting up with the scepticism and ridicule. He left finance in Hong Kong and exchanged his city flat for a hut in the Sumatran highlands, paying six dollars a month in rent. 'It should have been four, but I was white so they put the price up. I put in electric and water. I washed in a bucket. No one spoke English so I had to become fluent in Bahasa.'

Matthew recounts his story smoothly, almost as if he is giving a presentation to investors or speaking in an advert. There is little hesitation, little doubt, each detail fitting perfectly into the whole. In some ways his story reminded me of the Victorian quest genre; like

Furness, he embarked on a journey to retrieve an object from the wild, except the object in question, the excrement of a wild animal, was far stranger than anything Conan Doyle or Haggard could have concocted. 'It's almost slightly unfair to call it excrement, because when you have it fresh, it smells quite sweet,' he informed me. 'If you have it out in the wild it's almost like a semi-translucent gelatinous kind of syrup. I like to call them rough diamonds.'

Akin to a civet diamond mine, northern Sumatra was an astute choice of destination. The Dutch East India Company initially chose Java to be the centre of their coffee plantations, first importing arabica coffee in 1696 from India, but today it is Sumatra that is the heart of Indonesian coffee production. When Matthew arrived in the north of the island, he found that most of the coffee farmers in the Gayo highlands were only too keen to diversify their income and collect for him. 'They [the farmers] are poorer, relatively, than they ever have been,' he told me. 'Many of my guys had sold coffee to a guy who had then sold it to Starbucks. And what they were given for a kilo meant that they couldn't even afford the cheapest coffee. . . . There's nothing fair about that.'

When I visited Iceland, the eiderdown harvesters, aggrieved by their cut of the down's final price, often compared themselves to coffee farmers, the symbol of the inequities of global trade. Talking with Matthew, I could understand why. Desperately poor, Gayo's farmers struggled to make a living from the beans they sold on to middlemen. He saw his gourmet coffee as a means of changing this, offering them a decent price: 'I would make the Bugatti and then the Fiat 500. All of the DNA that makes the Bugatti could trickle down the pyramid. And then I could offer something not just to my farmers, but all the farmers, the rest of the farmers of the world, because I've proved a model that can trickle down to every level.'

Drawing on his academic background as a microbiologist, Matthew taught the farmers to source the civet scats, collecting this wild produce before it dried up. 'When the ambient temperature gets to thirty or late twenties, you're switching on all those stomach enzymes, reducing flavours,' he said. 'You have to get it fresh.' The farmers would head out at about five or six in the morning, hunting for wild civet excrement in specific places in their *kebun*, or farm. 'Civets are extremely territorial, so when they poop, they like certain places, fallen trees [for example].' Wary of spooking the civets, the farmers are careful not to take all the produce, leaving a bit behind. 'Otherwise the civet is like, "I'm a little bit disturbed but there's nothing here and my territory is gone."' I thought back to the harvesting techniques of old, the scraping, sweating, confinement and tail-pulling of London's captive civets. Matthew's harvest was different, less battery farming than foraging.

In the course of four years, Matthew sourced some four hundred kilos of civet scats from farmers in Sumatra. To process this coffee was a Sisyphean task; the coffee normally harvested by his farmers from their bushes is quite uniform, but beans within civet scats offer anything but consistency. 'The cat can eat any kind of fruit, any variety of coffee,' Matthew told me. 'I think a cat can circulate three to five kilometres in a radius from any one spot where it has pooped, and so it probably has a six-to-eight-hour digestive period.' Variations in density can cause serious problems during roasting. 'If you go and roast [beans of inconsistent density] you are going to get charcoal and green acrid [taste].' He spent years trying to develop a methodology, leaving beans from single farms in bags at altitude. 'So you leave the beans together in a bag for anywhere between three to six months, and they share water. All of a sudden your bell curve just tightens right up. In a cup of one hundred and forty [beans], I

potentially had fifteen outside of where I wanted them to be.' After
the roasting, he took ten kilos to Harrods, priced at £2,000 per kilo;
the store agreed to stock it. 'It doesn't have any bitterness,' Matthew
said proudly. 'It's rounded, it's silky, molasses and chocolate tones.
[After] you've drank it, it lingers.'

I n addition to eiderdown and edible birds' nests, among Ole Worm's
objects is a single coffee bean given to him by the renowned Dutch
collector Bernhard Paludanus on the occasion of the Dane's visit to
Enkhuizen in 1610. According to the Danish philologist Henrik
Schepelern, Worm's reference in his museum is the first actual description of a bean in Europe (although it was a German, Leonhard
Rauwolf, who in 1583 was the first European to relate the drinking
of coffee). At the time Worm received his bean, coffee was still a
curiosity in Europe, drunk by Sufi mystics in Yemen, Ottoman soldiers and officials, and traders. It had spread from Ethiopia to
Yemen, Cairo, Damascus and Mecca, but had yet to transition from
a gateway to the mystical to the global stimulant of capitalism sought
by merchants and writers including Daniel Defoe.

It is hard to reconcile Worm's bean with the coffee you might
find today, this mass product that is seemingly everywhere. Since
Worm's time, coffee production has spread all over the world: it is
now grown commercially on four continents and employs, from
bean to cup, millions of people. 'No longer primarily the beverage
of spiritual contemplation, commerce, or leisure, coffee became the
alarm clock that marked industrial time,' writes American historian
Steven Topik. Industrialised and consolidated, its trade is controlled
by a handful of commodity houses, such as ECOM, Neumann and
Volcafe. Today arabica beans form the bulk of world trade, but

increasingly it is robusta, a cheap, resistant cousin of arabica, that is gaining ground, particularly in so-called emerging markets. In this light one can understand the attractions of civet coffee. Like many gourmet coffees, it offered not just taste but the chance of distinction. In addition to a coffee bean, you had a wild animal, the civet – the champagne connoisseur picking the finest grapes.

Unlike Matthew, I did not feel drawn to the story of civet coffee, my own views perhaps closer to those of the cynical head of a trade association. Try as I might, I could not fall in love with the rough diamonds Matthew described or the descriptions of civets that I read. I felt his frustration at the homogeneity of coffee, but I did not see the need to generate or re-create the exotic, to participate in this arms race to produce something different for the market. I preferred the beans of Defoe's coffee house, this stimulant that brought people together, rather than creating distinctions through pricing. When I inspected my own civet coffee, I felt neither the bewilderment of birds' nests nor the idealism of eiderdown, only a sense of disgust. Yet there was a glint in these rough diamonds that I found alluring: unlike nests or down, the civet's scat had been properly discarded by the animal and could be gathered without disturbing its producer. I hoped that I might have another eiderdown in my cupboard, but without its obligations – an object that embodied a risk-free relationship based on the acts of tracking and collecting.

After I left the company of the birds' nest harvesters, I took a flight to the island of Bali, which, I had read, had become a showcase of civet coffee production. Over the past ten years, numerous plantations have opened offering tourists the chance to see how civet coffee is produced. Part of Indonesia, Bali is a short leap from Java, but it offers more points of contrast than similarity: it is majority Hindu, and its main exports are not palm oil or coffee, but tourism and rice. When I arrived in Ubud, I was greeted not by the call to prayer, but by the smell of incense and the tooting of taxis vying for the attention of tourists clasping Lonely Planet guides.

Bali's civet coffee plantations lay along one of the main roads between Ubud and Kintamani, advertised by a series of large bill-boards. At the first plantation I visited, I was greeted by the host, who led me in and introduced the different kinds of coffee grown on the island. Arabica coffee likely made its way to Bali from Ethiopia via Yemen, India and Java. A resilient plant, if left to its own devices it will shoot upwards to twenty feet in height, but it is normally pruned to a fraction of that. Its leaves are some six inches long, with white, fragrant, jasmine-like flowers and red fruits, akin to cherries, on which civets feast.

Ahead of me lay a battery of some six cages, populated by civets pacing back and forth hypnotically. This was a far cry from Matthew's story, transporting me instead to Defoe's house in Newington Green. It was not that they were probing the confines of the cage, exploring its corners; they were locked in a repetitive motion like a swinging pendulum. In some places the civets had gnawed through the metal wire. When one grasped the cage I saw that its feet were almost handlike, different from the pads of cats that I had expected. I bent forward to look more closely at them, but was repelled by the potent smell of the animal's gland.

As I inspected the captives, the guide related the practice of the civet coffee harvest. 'We feed them with banana and papaya,' he said, as if reluctantly reading a script. Although the civets were captive, they were released at night to gorge on nearby coffee bushes. 'We have them free in the jungle in the middle of the night time. They eat the coffee bean and then we collect the poo.' Like Matthew's, his story suggested a harmonious trade, one where the civets had time away from the cage, but it was not clear how they did not escape during the night. 'We have to cut their fingernails to make sure they cannot climb,' the guide assured me.

I looked around at each separate input into the coffee: the arabica plant in a pot, the haggard civets in a cage, the harvester relating his lines. It was hard to imagine them magically combining, the clipped civets roaming outside, gorging on the beans to produce the large quantity of uniform scats before me. When I pushed the guide on the origin of the beans, whether the harvest really took place, he broke down in tears. 'My boss, he just bought the coffee. We just process it to make the powder,' he said; the civets, meanwhile, never left the confines of their cages.

Originally from a village in the north of Bali, the guide said he had left his rice field and taken the job to pay for his son's education. 'I don't mind doing this for the support of my family. When I don't need much money for my life, I go back to the rice field,' he said. 'I don't want too much bullshit.' In his book *Bali: A Paradise Created*, the historian Adrian Vickers recounts how, in the twentieth century, European colonial officials, artists and tourists fashioned the image of Bali as an exotic paradise. Bali's civet coffee was no exception, a mirage conjured by its tourist industry. In reality there is little or no civet coffee production on the island; these plantations are but displays, theatres of the wild, for tourists. The true civet factories are elsewhere, beyond the gaze of tourists in Sumatra. 'You are not in the right place to learn about coffee,' the guide said. 'It's just bullshit to make the tourists come in, like you.'

When the tour of the civets came to a close, he offered me a menu with various choices of drink, including civet coffee, by far the most expensive. Like a client in a strip club, I had viewed the civets dancing for free, and it was now time to buy a drink. Feeling somewhat obliged, I chose a civet coffee, and drank a thick liquid that I found as tasteless as the instant coffee I routinely drank in the office. My head started to buzz with the heat, coffee and smell of civet, and I headed to another plantation, unsure whether I was documenting a trade or fuelling it.

n 1991 a British coffee trader named Tony Wild imported a single kilogram of *kopi luwak* as a curiosity, hoping it would elicit attention. After Wild publicised the sample, the local media picked up the story of 'crap', 'wolf' or 'cat' beans. Coffee companies began marketing the beans and their scatological narrative; other bizarre coffees began to appear in upmarket stores in London, beans that had passed through the digestive tracts of Thai elephants, Brazilian jacu (or guan) birds and bonobo apes. It was curious to think of the transformation of these excreted coffee beans. Once a symbol of identity for local farmers, civet coffee was now sipped at high-end coffee bars and discussed on *Oprah*. 'Tasting *kopi luwak* had become a not-to-be-missed life experience,' writes Wild, 'like seeing [Angkor Wat] by moonlight or attending Ascot Opening Day.' I thought of the history of offal, whose very name evokes discarded meat that 'falls off' a butcher's block. Once peasant food eaten only by the poor, it has been reinvented by the market as a gourmet dish, offered at steep prices in high-end restaurants.

In the eighteenth century, or perhaps earlier, Danish merchants began to export eiderdown, trading it in distant ports. The material, which once warmed only Icelanders, would have found itself in bedrooms all over the world, covering elites of different nationalities. In the case of eiderdown's expansion, little changed in the Icelandic technique of harvesting: the lives of Iceland's ducks continued even as the bodies it warmed changed. The history of civet coffee was different. As soon as it crossed borders, shedding its local garb, Indonesian farmers started farming civets to increase production. There was no enterprising development in harvesting techniques, but a strange regression to the practices of Defoe.

When Wild learned about the conditions of this new industry, he felt bad. 'I feel as if long ago I must have inadvertently put my finger on the pulse of some monstrous zeitgeist, a grotesque cancer that constantly mutates into yet more vile and virulent forms,' he wrote. Feeling responsible, he launched a campaign called 'Cut the Crap', publicising the iniquities of the civet coffee trade, and made a documentary with the BBC about it, travelling to Sumatra to expose civet farmers. He had a degree of success: Harrods pulled civet coffee from its shelves, with the exception of Matthew's own wild-harvested brand. But there continues a lively trade in civet coffee, supplying demand from China and Indonesia as well as from European and American travellers like me. That much was evident when I passed through a WHSmith, the British retailer, at Bali's airport and found stands devoted to tape, ballpoint pens, chocolate bars and civet coffee.

Despite the efforts of Wild and other campaigners, I could not imagine civet coffee disappearing anytime soon. Like so many illicit commodities, from cocaine to opium, the growth in civet coffee is correlated to fluctuations in the price of staple crops. Since the 2000s the price of Sumatran coffee has plummeted, a fall often attributed to the rise of coffee production in Vietnam funded by the World Bank and to the disappearance of coffee marketing boards. It is, Matthew says, no coincidence that civet coffee production has boomed during this period, as farmers look to hedge against volatile prices. 'I don't agree with animals in cages in any way, shape or form,' Matthew told me. 'When I drank some [coffee from caged civets] I was sick for three days. It was horrendous. [But] you've got to look at the bigger picture and say, Why is this industry out there? It's the easiest way for people to make money who are being screwed by the system.'

L ike so many of his schemes, Defoe's civet production facility never took off. One day in mid-October 1692, two sheriffs of London or their agents rode out to Newington and seized the 'seventy guiltless civet-cats' on behalf of His Majesty King William III. It transpired that Defoe had never actually purchased the civets, paying less than a quarter of the price and getting the rest on credit. He then persuaded his mother-in-law, Joan Tuffley, to pay the creditor off, until another creditor, Sir Thomas Estcourt, tried to take ownership of the civets; the creatures became the subject of a legal dispute between Defoe and his creditors, including his bewildered mother-in-law.

Defoe moved on, declared himself bankrupt, submitted himself to London's Fleet Prison, fled Britain, and later wrote *Robinson Crusoe*. The civets, for their part, were advertised for sale again, no doubt to be scraped by someone else. In later years literary critics and historians reflected on the meaning of Defoe's encounter with the civets. What did the civet dispute say about Defoe's financial probity, his relationship with his creditors? Was he honest in his dealings? The Canadian scholar Theodore Newton saw him as a sort of George Parker, the con man who managed to 'sell' public landmarks like the Brooklyn Bridge to unsuspecting immigrants.

'He [Defoe] worked the modern "Statue of Liberty" confidence game two centuries before its time,' Newton concluded in 1937, 'for he knowingly aided others in selling something which was not his to sell.'

In all this analysis of the Newington Green affair, little regard is paid to the civets, the motors of the perfume trade. Reduced to the status of collateral, the creatures are discussed as if they were an inanimate factor of production, a litmus test of human relations. Much has changed today. After Wild's campaign, the civets themselves became the centre of the discourse about civet coffee, their plight discussed and analysed, their conditions unpicked, their behaviour measured. By contrast, the harvesters appeared a monolithic entity, a symbol of our broken relationship with the wild, embodying a thoughtless cruelty. I sought to speak with civet harvesters, but was met with silence. After Wild's investigation, producers in Indonesia went to ground, shunning interviews; even producers who advertised wild *kopi luwak* did not want to talk, with the exception of Matthew, but even he expressed little interest in showing me his harvest. The production of civet coffee remained elusive, a subject of myths, disinformation and falsehoods.

Indeed, the greatest of these turns out to be the foundational myth of civet coffee, the image of the civet as the champagne maker, selecting the best grapes. When I got back home, I read that early Dutch botanists, concerned about the spread of borer beetles, had examined civet scats to see what sort of beans they ate. Far from selecting the 'finest' cherries, the civets had eaten a significant number of beans with larvae or beetles, which had survived the process of digestion. As Colin Cahill, an American anthropologist, writes, 'While civets might eat the ripest, best cherries, "best" for a civet might not always mean "best" for the coffee industry.'

*

Somewhere buried in all this storytelling were the civets. Kidnapped, scraped, traded and exhibited, they sat at the bottom of this chain, bearing the weight of the world's desire for the strange. After so much time inspecting them in cages, I struggled to imagine them surrounded by tree branches rather than chicken wire, gorging on the forest instead of on a uniform diet of coffee cherries. For a moment I wanted to glimpse them removed from our own desires and projections, before they entered the production line of coffee or perfume. I headed for Cipaganti, a village that sits adjacent to Mount Papandayan, an active volcano in western Java.

The fate of South Asia's forest dwellers is closely tied to the trees in which they make their home. As Java's forests have disappeared, so too have the slow loris, a small primate that curls up into a ball on its branch; the Javan mouse deer, the smallest known ungulate; and the Javan rhinoceros. The civet has fared better, adapting to human structures from telephone wires to irrigation hoses, although civets are regularly trapped and killed because locals consider them to be pests. Among the last refuges of arboreal creatures are the slopes of Java's volcanoes and mountains: often too steep for intensive cultivation, they act like small arks, as the tides of agriculture and development rise and rise.

The mountains surrounding Mount Papandayan are no exception, their slopes home to leopard cats, slow lorises and civets. Their numbers are closely monitored by the Little Fireface Project, a small conservation initiative based in Cipaganti, dedicated to understanding and protecting the slow loris, a critically endangered primate. About half the size of a tree sloth, the slow loris has large eyes that dominate its small face and long, powerful arms. It is a strikingly strange and beautiful creature, but it is cursed by its beauty. Highly sought after as pets, lorises are routinely plucked from trees and sold to traders.

Those lorises that remain in Cipaganti are monitored closely, named and tagged by the project's small team of biologists and trackers. One cool evening I joined them as they headed out into the forest on the slopes to track Fernando, a male slow loris, by following the signal emitted by his radio collar. Known by the team as a local lothario, Fernando was a young, energetic loris, prone to range widely in search of females or mischief. It was tiring work tracking him, clambering through fields of tea, coffee and banana. Despite their name, slow lorises are surprisingly nimble, using their long arms and strong hands to traverse branches in the search for the insects, nectar, gum and fruit on which they feed, and we had to exert ourselves to keep up with Fernando. I strained my ears, but noise seemed to emanate from every tree, refracted as if in an echo chamber.

While Aconk, one of the local guides, tracked the slow loris, Robert O'Hagan, the project's coordinator, showed me how to spot civets. Civets prowl like cats, creeping along tree branches or narrow irrigation hoses, foraging for fruits and berries. Instead of using a radio receiver, Rob and I relied on sight, scanning the trees and hoses with the red beams of our headlamps to pick up reflections from their eyes. Much like cats, civets possess *tapetum lucidum*, or 'bright tapestry', a layer of tissue in their eyes that increases the sensitivity of their vision. Unfortunately for the uninitiated, they share this adaptation with many other animals in the forest, from leopard cats to slow lorises, ferret badgers and a host of domesticated animals.

Like jewellers, Rob and the trackers had learned to read the reflective glints of eyes, these precious stones of the forest. Almost instinctively, they noted differences in intensity, size and the rate of blinking, and narrowed down the list of potential owners, but I remained lost, unsure whether I was looking at a precious stone or pyrite. In the distance I made out a pair of glints at knee level, possibly a civet or a wild pig. 'Oh,' Rob said impassively, 'that's a dog.' The only traces of the civets that I could find were their small scats. Unlike the eiders' glinting quartz, these droppings were elusive, rare finds in the grass. I thought of Matthew's story, his quest to gather these rough diamonds. Did he really manage to harvest four hundred kilos of these scats? Or was his story just another fiction? How did he know they were from the wild?

'There's visual markers, there's aroma markers,' Matthew had told me confidently. 'One of the marks is [that] when a civet cat takes a coffee cherry off a tree, it normally takes a bit of wood that the cherry is fixed to. So you'll find small pieces of wood. If it's caged, a human has taken the wood off. You also have variability of seeds and stones from nut and fruit skins; they eat ravenously and sometimes they don't digest their skins.' I looked down at one of the scats below me. Unlike the scats I had seen in Bali, these were chaotic and rough-edged, full of different seeds and twigs, rather than a single type of bean. Documents of the forest, their scats spoke of its varied vegetation, but they were more than mere representations; as the civets gorged on the fruit of these slopes, they shaped them, too, helping to sculpt an entire ecosystem. Perhaps Matthew's story was one truth in this industry based on fabulation.

Towards midnight, Aconk, shivering in the cool air, picked up the signal of a slow loris with his receiver. We shone our headlamps upwards and caught its large eyes high on a branch, to which it clung, koala-like. Unlike the civet, it looked so delicate, so frail, so domesticated, but Rob warned me that it could be quite dangerous. If threatened, the slow loris may lick a small gland under its armpit, which secretes a toxin. When mixed with the animal's saliva and delivered with a strong bite, it can be deadly. Those who illegally trade the slow loris get around this by removing its teeth, an operation that may kill or cause serious harm to the creature.

As you proceed higher up from Cipaganti, the sound of the call to prayer diminishes to a faint wail, the rasps of motorcycles fade to a distant hum, smothered by the calls of crickets. The plantations of tea and chayote gourds drop away, replaced by areas of primary forest too steep to be cultivated. It was around this borderland that we saw a figure on a narrow branch, a set of glinting eyes up in the trees, which Rob identified as an Asian palm civet, *Paradoxurus hermaphroditus*. It was not the specific glint that alerted us to its identity, but its position, traversing the tightrope of the tree branch, its long tail extended, affording it balance like a fifth limb. It turned its head towards us, seemingly uninterested, and continued its tightrope act into the darkness.

Admiring the civet in the tree, I was reminded of those early zoologists who were unable to conceive of a creature as new or separate. The palm civet nimbly traversed the branch like a domestic cat, but its features were sharper than those of any feline, its pointy snout and long, slender body reminding me of a stoat, weasel or mink. From a distance the civet seemed a creature of burrows, dens or warrens, a digger rather than a tree dweller. Had our headlights been stronger and stripped of their red lenses, we would have made out its striped fur, similar to that of a hyena or skunk. When I visited Bali, the civet had struck me as a fierce beast, akin to a rat or ferret, but now it seemed a different creature. Surrounded by the forest, its lithe form at home among the limbs of the tree, it had a certain feral elegance, proceeding along a natural line rather than pacing the confines of a cage. Civets were prized as strange, wild, unclassifiable creatures, but they did not seem strange here, their sweet smell part of the forest's aroma. No, it was we who had made them strange, tearing them out of their fabric, installing them in a new architecture, arguing over their origins, and selling their excretions or secretions for profit.

SEA SILK

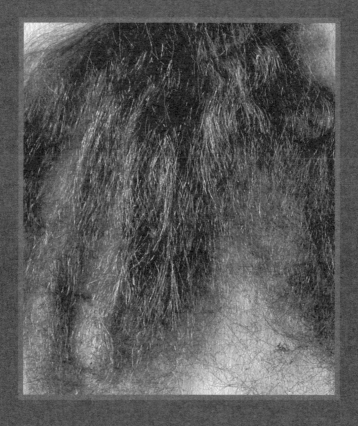

n 2001 the British artist Michael Landy set out to destroy all his worldly possessions. He teamed up with Artangel, the UK arts organisation, and began compiling an inventory of all the objects he had accumulated in his life. All in all, he documented some 7,227 items, including his father's sheepskin coat, love letters, his Saab 900 and a roll of toilet paper. After displaying the objects on a hundred-metre conveyor belt in a former clothing store in London's Oxford Circus, he and twelve assistants proceeded to destroy all of them in front of bewildered shoppers. 'In one sense, the message was: where are we heading?' he told the art critic Alastair Sooke. 'The more stuff people have, the more successful we perceive them to be – but if we all end up with 7,227 things, then we won't have a planet.'

I read about Landy's show, titled *Break Down*, while still at school. It made a strong impression on me because it offered a quick path to reinvent one's teenage self. I'd periodically clear out all my belongings, only to find that I remained little different, possessed of the same habits and insecurities. It was a futile act of rebellion, but now, approaching Landy's age when he carried out the show, I viewed *Break Down* differently. If we could truly know the origin of the objects that surround us, acquaint ourselves with the lives that they have touched, the landscapes they have shaped, would we be able to keep them? After returning from Bali, I promptly threw my civet coffee away, wanting to disentangle myself from its stories of

cages and cruelty. The presence of this object led me to look sceptically at others in my own possession: What other secrets lurked in
my drawers or on my shelves?

Since the 1990s, a school of free market economics, calling itself
enviro-capitalism, has argued that an effective way to conserve nature
is to monetise it. In this light, the trade in eiderdown, birds' nests and
even civet coffee could be seen as a great success: protected by harvesters who are incentivised by profit, the numbers of the source species had swelled. But such a view captured only a partial truth, one
that ignored the destruction of eiders' predators, the ransacking of
caves or the ill treatment of civets. After recently spending time with
caged animals, I was overcome by a desire to seal off these creatures
from the market, to erect barriers that might protect them from an
encounter that was inevitably traumatic. I wanted to learn from a material that had a special place apart from our own lives, one that would
never be found in a supermarket or Amazon warehouse, one that
could never be thrown away, because it could never be owned.

I n a small apartment in Cagliari, the capital of Sardinia, an octo-
genarian woman handed me a box full of twenty or thirty locks of
what appeared to be dark human hair. In my hand the threads felt
as weightless as silk, but it was their curious shine that drew my attention. 'The golden-brown sheen is said to be unique among the
fabrics, soft as vicuna wool, bright, very slight,' enthused a French
writer in 1806. I picked up one of the locks and held it between my
thumb and index finger so that it could catch the light. At the lock's
base it had coalesced into a dull, matted, leathery material, but
gradually it split up into hundreds and hundreds of hairs, dark
brown to a shining gold.

In the eighteenth century an English traveller, Henry Swinburne, likened the colour of such locks to the 'burnished gold on the back of some flies and beetles.' Although revealing, Swinburne's comparison gives little clue as to the origin of the threads; no insect was their author. More helpful is an essay that appeared in AD 209, written by Tertullian, an early Christian writer from Carthage, on dress in Roman society. After discussing conventional clothing materials, such as wool and flax, he makes a curious reference to 'fleeces' that were 'obtained from the sea'. 'Nor was it enough [in the Roman Empire] to comb and sow the material for a tunic,' he writes. 'It was necessary also to fish for one's dress.'

When we think of silk we often conjure images of insects: silk-worms, spiders, beetles or wasps. But there are also, to quote one French Enlightenment thinker, *'coquillages à soye'*, that is, silk shells: bivalve molluscs, such as mussels, clams, scallops, and pen shells, that produce silken threads known as byssus. Unlike insects that use their silk to build cocoons, molluscs employ byssus as an anchor, tethering themselves to the sea floor against the push and pull of waves. 'All sea creatures that do not swim or that swim with difficulty fear the motion of the element that surrounds them,' explained René Antoine Ferchault de Réaumur, the French entomologist who first untangled the mechanism by which mussels produce byssus.

While his contemporaries tested the properties of edible birds' nests, Réaumur took a different approach with this aquatic adhesive, observing how mussels produce silk in real time. Much like present-day biologists, Réaumur placed mussels in a saltwater tank and waited to 'catch them in the act'. Before long, he observed a small tongue emerge from each of the mussels. 'I saw them feel out their tips to the right, left, rear, front, as if to explore their surroundings,' he wrote. On finding a solid surface, the tongues would pause and then suddenly retract, leaving behind a fine thread. Over hours and days, the tongues repeated this action, gradually building a network of threads between the shell and the solid base. '[The organ] is not an arm, nor is it as a leg that we must regard it; it rarely carries out these functions. We must view it as a spinner.'

Réaumur was intrigued by the way that creatures produced materials to aid their survival, and he often hoped to replicate their techniques in manufacturing. By the time he examined these mussels, he had already written extensively about the production of silk by spiders and other insects. 'Spiders and caterpillars . . . ', he noted, 'form threads of any required length by making the viscous liquid of which the filament is formed pass through fine perforations in an organ appointed for spinning.' Did mussels do the same, he wondered, pulling silk out like a 'wire drawer'? Cutting open a mussel, he observed that a small gland produced a liquid that flowed into a channel in the tongue, solidified, and was then released to form a thread. If silkworms 'drew' their silk, mussels, he concluded, moulded it as 'the founder who casts metal'.

I thought back to all the plastic I had found in seemingly remote places, from the pastor's dildo to the bottles strewn on Mantanani, all those objects moulded to fit our own desires and needs. Were Réaumur alive today, perhaps he might have compared byssus formation to injection moulding, where molten materials, such as plastic, are injected into a mould and left to cool, assuming the form of the cavity. But unlike injection or cast moulding, byssus does not require a change in temperature to set. On contact with water, the amino acids present in byssus harden to form a solid anchor. While many adhesives – from the mucus of a swiftlet's nest to synthetic glues – often dissolve in water, that of mussels remains firm, eliciting hope among chemists that a synthetic glue may one day be able to do the same, resolving a host of mechanical challenges in the medical field.

The particular strands that I held were much larger than the mussel byssus examined by Réaumur. Moulded by the *Pinna nobilis*, or noble pen shell, a giant bivalve cousin of the common mussel, they measured about twenty centimetres in length, testimony to the huge size of the pinna's tongue, or 'spinner'. Found throughout the shallows of the Mediterranean, from the Greek to the Croatian, Tunisian and Spanish coastlines, the *Pinna nobilis* perches in the sand, emitting hundreds of byssus strands as an anchor. Floating above these great shells and looking downwards, I was reminded of gnarled tombstones, their clean-cut angular forms masked by giant green weeds. Some shells, often encrusted with barnacles, seaweed and coral, live up to twenty years and can reach more than a metre in height.

Ancient writers were intrigued by the *Pinna nobilis*, silently filtering the water of its nutrients. In the fourth century BC, Aristotle described the shells that 'grow upright out of the bottom in sandy and slimy places.' Noting that the pinna was rooted to the sea floor, he argued that 'broadly speaking, the entire genus of testaceans have a resemblance to vegetables, if they be contrasted with such animals as are capable of progression.' Such a view of molluscs remained largely unchallenged until the eighteenth century, when the Italian polymath conchologist Giuseppe Saverio Poli studied the soft parts of molluscs, revealing the complexities of their anatomy and nervous system. Like students of human anatomy, Poli took moulds of molluscs, including the pinna, into which he later poured hot wax and even planted a real lock of byssus. Admiring images of this mould today, I was struck by the complexity of the mollusc's organs, its eyes, heart, gills and liver, and the strange dark hair that sprouted forth from its pink flesh.

In their raw state underwater, the beards of the pinna resemble brown moss, full of algae and small shells, but once cleaned and combed, they are transformed into golden threads, commonly known as sea silk. Since at least the time of Tertullian, nearly two thousand years ago, these threads have been prized for their shine and strangeness. The scholar Felicitas Maeder of the Natural History Museum of Basel in Switzerland has carefully scoured written sources, hunting for references to sea silk in ancient to modern texts. The evidence is patchy, she says, but she has compiled an intriguing number of possible references to the fibre, including a mention of 'a fine cloth which some people say is made from the down of "water sheep"', in the *Hou Hanshu*, a Chinese source relating the history of the later Han dynasty, of the first to third century AD. According to the sixth-century historian Procopius, the Byzantine emperor Justinian I gave a gift of sea silk to five Armenian satraps: 'a cloak made of wool, not such as is produced by sheep, but gathered from the sea.' Unlike textual references, archaeological evidence for ancient sea silk production is notably scarce; the oldest sea silk object dates to the fourth century: a cloth fragment found at the site of Aquincum, a Roman legionary town, now in Budapest.

What is certain is that by modern times, sea silk starts to emerge consistently in the historical record, appearing in texts, cabinets of curiosity, private collections and exhibitions. Associated with decadence and rarity, it was sometimes mixed with other materials, possibly including vicuña fibre. Far from being confined to exhibitions, sea silk starts to occupy a space in literature, entering the realm of the imagination. Young men, often Englishmen, mention sea silk in accounts of their travels to Italy as part of the Grand Tour. In 1804 the British admiral Horatio Nelson wrote to his lover Emma Hamilton that he was sending her 'a pair of curious gloves, they are made

only in Sardinia of the beards of mussels [*sic*]. I have ordered a muff;
they tell me they are very scarce, and for that reason I wish you to
have them.' Jules Verne, clearly intrigued by the qualities of this
strange marine fibre, chose to dress his narrator in *Twenty Thousand
Leagues Under the Sea*, Dr Pierre Aronnax, in 'sea-boots, an otterskin
cap, a greatcoat of byssus lined with sealskin'.

The history of silk is largely one of consolidation and industrialisation. Domesticated some five thousand years ago, the silkworm –
Bombyx mori – is now bred intensively all over the world, most
notably in China and India. The history of sea silk could not be
more different. In the 1930s in Taranto, a small coastal city on the
heel of Italy, a marine biologist named Attilio Cerruti did experiment with pinna aquaculture, but it proved too labour-intensive and
time-consuming to plant the larvae of pen shells as if they were
seeds. The most common means of harvesting was to rely on the
traditional methods, where fishermen would haul the shells up using metal tongs, an exhausting process, and remove the byssus by
hand. Then there was the cleaning, spinning and weaving of byssus,
a repetitive process often carried out by women. Harsh and full of
impurities, byssus had to be painstakingly worked, again by hand. In
1916, Giuseppe Basso-Arnoux, a Sardinian doctor, related how he
sent a batch of sea silk to a textile factory near Basel, but it was rejected on the grounds that it would take too long to clean. According
to Maeder, sea silk weaving remained a local practice, confined
mainly to the home, girls' schools, orphanages and convents.

When Tom Harrisson made 'Birds' Nest Soup', he was aware
that the birds' nest trade was slowly dying. 'What is visibly
decreasing . . . ', he lamented in 1960, 'is a body of ultra-able-bodied
men prepared to run the deadly risks incurred in this singular and
literally precarious method of earning a living.' The history of sea

silk in the twentieth century proved similar, though its limiting factor was not deadly risk, but the huge amount of labour required to transform these strands into usable fibre. 'The women of that land [Sardinia] are unwilling to undertake such patient work,' complained Basso-Arnoux in 1916 after he abandoned his attempt to start a byssus industry. By the latter half of the twentieth century, knowledge of sea silk weaving was probably confined to a handful of women in Taranto and Sant'Antioco, a small island off the coast of Sardinia. With the passing of each decade, the number of women who possessed such knowledge grew smaller and smaller.

Reading about sea silk, I felt an affection for this strange material, whose own story ran so counter to the stable rhythms of eiderdown or the frantic gyrations of birds' nests. Countless individuals, both men and women, had sought to turn sea silk into an item of trade, celebrating its magical properties, but it had not found a place in our own century, its stories often ignored, its threads confined to museum basements. Harrisson's response to the decline in the birds' nest trade was to document it, to carefully record its details for posterity; I wanted to do the same for sea silk, but this time record the voices of those who worked it. I started to look for weavers who might know how to do so, scouring newspapers, books and scholarly studies, but I was met with disappointment. Margherita Sitzia, a sea silk weaver from Sant'Antioco, had died in 2011 at the age of ninety-six; another weaver, Efisia Murroni, reportedly passed away in 2013 at one hundred one.

I began to lose heart, but pleasingly all was not lost: there was, I read, one weaver left who knew how to work sea silk: Chiara Vigo, the last maestro of byssus. Then in her mid-sixties, she lived on Sant'Antioco, which is connected to Sardinia by an isthmus built by the Phoenicians. I read that at dawn and dusk each day she descended to the water to wail in ancient tongues and acknowledge those of her ancestors who had passed the secrets of sea silk down the generations. She dived down into Sant'Antioco's lagoon, and trimmed the beards of molluscs, then cleaned, spun and wove their golden fibre to produce beautiful tapestries and articles of clothing, which she displayed in her small museum in the island's main town, which shared its name.

n December 1938, Marjorie Courtenay-Latimer, a thirty-one-year-old South African museum curator, received a phone call from the docks of the city of East London on the Eastern Cape. An avid collector of natural history specimens, she had asked fishermen to let her know if they ever came across any unusual fish. When she arrived at the docks, she was not disappointed by the organism she found. 'I picked away at the layers of slime to reveal the most beautiful fish I had ever seen,' she recalled. 'It was five foot long, a pale, mauvy blue with faint flecks of whitish spots; it had an iridescent silver-blue-green sheen all over. It was covered in hard scales, and it had four limblike fins and a strange puppy-dog tail.'

The specimen, it was later established, had a number of characteristics that set it apart from any other living fish, including an oil-filled gas bladder and incompletely formed vertebrae. The chairman of the East London Museum initially dismissed it as a 'rock cod', but when Courtenay-Latimer shared the specimen with her friend James Smith, who taught at Rhodes University, he recognised it instantly as a coelacanth, a fish thought to have gone extinct millions of years ago and believed by some to be the direct ancestor of humankind. 'There was not a shadow of doubt,' Smith recalled. 'Scale by scale, bone by bone, fin by fin, it was a true coelacanth. It could have been one of those creatures of 200 million years ago come alive again.'

When writers and filmmakers entered Chiara Vigo's museum, they described her in similar terms to the coelacanth, this living organism previously thought to have gone extinct in the distant past. Like a shaman of old, she harked back to an era long before our relationship with nature was broken by capitalism, a time when, according to John Berger, 'animals constituted the first circle of what surrounded man'. Vigo saw the *Pinna nobilis*, unlike Sardinia's forests, fish or sheep, as more than the prospect of food, fabric or money; the mollusc was sacred, magical and oracular. As Sardinia struggled through the post-war period, its farmers heading off to work in factories and the island turning to mass tourism, Chiara and her ancestors had conserved her traditions in her museum.

Intrigued by the accounts about Vigo, I rented a car in Cagliari and drove along the island's southern coast to meet the last maestro of byssus. The coastal road winds past magnesium-white beaches, sixteenth-century towers and the island's huge oil refinery, Saras, one of the largest employers in Sardinia. It's a much longer drive than the direct route inland, but it is one of the most breathtaking in Sardinia, offering the island's characteristic mix of ruined guard towers and monumental industrial chimneys, many of which were built during the island's drive toward industrialisation in the 1960s. That day I drove carelessly, my gaze constantly drawn towards the beach, and it was with some relief that I approached Sant'Antioco.

Writers, travellers and tourists who visit Sant'Antioco often comment that it has a fairy-tale quality, as if its isthmus acted as a bridge across time as well as water. As in so much of Sardinia, traces of its ancient history lie close to the surface: Stone Age menhirs,

Carthaginian tombs and Roman aqueducts dot the island. As I crossed the isthmus, my attention was drawn to the pink flamingos and fishing boats taking shelter in its shallow lagoons, shielded from the powerful wind, the *maestrale*, which blows from the north-west. I saw countless locals picking molluscs, bent over in the calm turquoise water as if harvesting rice, ignoring the tourists speeding by in rental cars on their way to the island's beaches, fish restaurants and museums.

Many of these tourists would be heading for Chiara's museum, which lies in an old palazzo a short walk up a hill in town. Built in 1820, it is a handsome, spacious building once used to store the town's grain. In recent years it has become one of the island's big attractions, drawing thousands of visitors annually. As I walked towards it, I found a large group of tourists milling around at the entrance, shepherded by a local guide. Looming over them was a large sign depicting a woman holding a spindle, head bowed and eyes closed as if performing an act of penitence. 'The Museum of Byssus', it read. 'Maestro Chiara Vigo.'

I found the weaver seated at a desk, her head inclined towards a clump of byssus, a spindle and a beaker with a yellow liquid. Chiara wore a blue jumper and striking red glasses and had long, perfect fingernails. All around her were countless curious objects, plucked from the sea or her own home: beautiful sea silk embroideries, letters from visitors, photographs of Chiara by the sea, a papier-mâché lighthouse, several paper pom-poms, an oversize metal *Pinna nobilis*, various additional beakers, framed excerpts from ancient texts, a certificate signed by Silvio Berlusconi, a loom, a diorama with shells and nets, and a wooden offerings box. It was high season in the Byssus Museum, and tourists and guides were piling in its double doors, eager to catch a glimpse of the weaver and her art. The stone floor cracked with the sound of a pair of high-heeled cream-coloured shoes belonging to the maestro's elegant assistant, her ear glued to her phone. 'No,' the assistant said sharply down the line, 'the maestro is not available. She is fully booked. Call back tomorrow.'

Many tales have been told about sea silk. In the second and third centuries, Chinese traders spoke of aquatic sheep, which would emerge from the sea and scratch against rocks, leaving their wool behind. Arab traders in the tenth century told of a cloth that was made from the hairs shed by a sea monster called *abu qalamun*. Chiara, too, added her own stories to this history. The Chaldeans, Egyptians, Greeks and Jews all used byssus to make their sacred vestments, she said; there were even references to it in the Bible. 'Remember when it says that King Solomon appears "shining" in public? Why do you think that is?' she asked. 'He was wearing byssus-made clothes, that in the dark appear brown, but once in the light, they shine like gold.'

After recounting her ancient history of byssus, the maestro explained how she cleans, dyes and weaves the fibres. 'Once collected, the sea silk is desalinated for twenty-five days, with fresh water replaced at least every three hours, day and night,' she said. It is then left to dry and brushed to remove bits of seaweed and shells. She took a clump of byssus, brushed it, and put it in my hand. 'What do you feel?' she asked. Although byssus is often described as finer than silk, this is not always strictly true: its diameter can vary greatly, between ten and fifty microns, and it is comparable to other natural fibres, such as mulberry silk. My eyes closed, I did feel a slight brush against my hand, and yet I did not have the courage to say any of this to Chiara, who, in any case, was quickly moving on to pass the byssus to another visitor.

*

n 1777 a German clergyman and conchologist, Johann Chemnitz, wrote of the fantastic quantities of pen shells in the Mediterranean, describing outcrops of the shells as 'cities and villages in the depth of the sea where the oldest members protrude like towers'. But in the twentieth century their numbers began to decline. Fishermen's anchors destroyed the molluscs' own fibrous moorings; trawler nets sent the shells tumbling; and collectors plucked them from the sand as souvenirs. If you find a pen shell in a meadow of seagrass today, it may be the only one within an area of one hundred square metres, more solitary tower than conurbation. In 1992 the Italian government declared the *Pinna nobilis* endangered, and banned its fishing, ostensibly signalling an end to sea silk weaving.

Two decades after the introduction of this legislation, an Italian cruise ship, the *Costa Concordia*, ran aground off the western coast of Italy. During operations to salvage the wreck, divers discovered a forest of some two hundred *Pinna nobilis*. Fearful that the shells might be harmed in the course of engineering work, divers moved the shells to a site several hundred metres away, where they were planted in neat rows. The German magazine *Der Spiegel* likened the transferred molluscs to a cemetery.

Observing the maestro and her supply of byssus, I found it curious that she had such a large supply of the threads, given that the *Pinna nobilis* was protected. When asked about this, she revealed that she had devised a special way to harvest sea silk without harming the molluscs. During the summer months, from the first full moon of May, she would descend to the sea, swim down to the shells, and cut off part of the mollusks' beards. It was exhausting, repetitive work; she said she had to undertake a hundred dives to harvest three hundred grams of beard, which would become thirty grams of cleaned material.

At the height of sea silk production in the eighteenth century, Italian fishermen would haul the molluscs onto their boats using various traps, tongs and forks. 'The Pinna is torn off the rocks with hooks, and broken for the sake of its bunch of silk called Lanapenna,' recorded Henry Swinburne in the nineteenth century. After opening the shells, fishermen would cut off the beards and scoop out the flesh, pieces the size of rump steaks. The maestro's method seemed so different from this practice; she foraged the sea silk instead of felling its owner. Without waiting for an answer to her own question about the fibre's weight, Chiara took the tuft back from another visitor, pulled apart some of the fibres, and attached them to the hook of a wooden drop spindle. 'Watch me make a thread with my magic wand,' she said mischievously. As she set the spindle spinning, the fragmented byssus twisted into a single thread. Effortlessly she fed more and more fibres, the thread lengthening from her fingers as if it were a miniature stream of dark golden liquid.

To understand the history of byssus in Sant'Antioco, Chiara said, you need to go back to the first century AD, when Berenice, the daughter of King Herod Agrippa I, arrived there. The first maestro (or custodian of the secrets of byssus), she passed her knowledge on to her daughter or granddaughter. Over the centuries, the secrets passed down through the generations until they reached a master weaver, Maria Maddalena Mereu, who in turn passed them on to her granddaughter Chiara Vigo, who swore the sacred oath of byssus. 'The old maestro leaves the water,' she said, '[and] gives a fistful of mud to the new maestro. The new maestro washes herself [and] dives in the water – from that moment that passage is made.'

The word *maestro*, derived from the Latin *magister*, is often used in Italian to denote a highly skilled artisan, for example *maestro falegname*, a master woodworker. We often think of a maestro as a teacher, and yet Chiara's definition of the word was different. Often talked about but never precisely defined, her term seemed to encompass conservation, craft, mysticism and a hostility to markets. 'It's a little bit like kung fu,' one of her students, an Australian artist, clarified. More than an ability to work a material, the status of maestro appeared to confer certain rights, that is, to harvest, spin and weave byssus. Unlike crafts that offered apprenticeships, the right to byssus was solely hereditary, transmitted from generation to generation. Indeed, Chiara said that the only person who might learn about the fibre's secrets was her daughter, Maddalena, then a university student in Dublin. 'In Sardinia knowledge of the arts or medicine often passes down a family line,' said Giuseppe Mongiello, an Italian artist who studied under Chiara with about ten others. 'You also find this in other societies in Africa and Asia.' Mongiello had worked in the museum since 2012 but had never touched byssus, only other materials like wool. 'She can't teach us. Because of the oath.'

Anthropologists have long lamented the loss of traditions such as myths, dances and performances, in the face of industrialisation, migration to cities and environmental degradation. In the 1960s planners in Rome decided to place a large bet that the future of Sardinia – an island of miners, shepherds, and farmers – lay in industrial production. Countless petrochemical plants, factories and refineries were built throughout the island as part of the state-led

Piano di Rinascita (Plan for Rebirth), transforming the entire island. 'No more ploughs to carve the earth, no more oxen,' wrote Bachisio Bandinu and Salvatore Cubeddu, two Sardinian intellectuals. 'The constant production of goods replaced the uncertainty of the harvest.' Reading about this history, the scale of change in Sardinian society, I marvelled at how Chiara's oral history of byssus had survived; the sacred chain between her and Berenice remained unbroken. Perhaps this explained why, according to the Italian press, Chiara's work had been proposed in 2005 as a candidate for UNESCO's Intangible Cultural Heritage list, defined as 'traditions or living expressions inherited from our ancestors and passed on to our descendants'.

Procopius relates that Justinian I sent 'certain monks' to Central Asia who smuggled silkworm eggs to the Byzantine Empire to form a new industry, breaking Chinese control of the trade. Not unlike many a shaman, witch doctor or silkworm breeder, Chiara fiercely protected her knowledge, refusing to disclose the details of her techniques for harvesting or working byssus to prying researchers, scientists or journalists. 'I have never published a study, because it is dangerous for the sea,' she said. 'All those who are in design, fashion, they will come here and carry out a massacre [of the pen shells].' Even Giuseppe, her closest student, was not permitted to accompany Chiara on her foraging trips. 'She goes alone at night when there are *bonacce*,' he told me. 'When the water is calm.'

Yet Chiara did permit us a small glimpse of one of the secrets she received from her grandmother: the means by which byssus can be lightened, the formula for converting the dark brown fibre into a golden thread. 'It's my family's recipe, more than six thousand years old,' she explained. After unwrapping the threads from the spindle, she dunked them into a glass beaker full of a yellow liquid. 'It has fifteen algae and two types of lemon juice.' Wailing into the beaker, she produced a sound similar to the blow of a conch horn. She raised her head as if in a trance and removed the threads from the beaker, then squeezed them to purge the yellow liquid. 'You want to see it become gold and light?' she asked. Without waiting for an answer, she got up and walked around the audience, displaying the golden thread. A beautiful light gold colour, it stretched as she pulled it with her fingers. 'It's like elastic,' someone whispered, enchanted.

Sea silk has always been highly prized, its bright shine and curious origin attracting the attention of merchants, the aristocracy and collectors, many of whom were willing to pay a high price for the golden threads. In the eighteenth century, Giuseppe Capecelatro, the archbishop of Taranto, became interested in the commercial potential of byssus. An ardent social reformer, he believed that the creation of a sea silk industry might improve the condition of Taranto's poor. He lost no opportunity to market byssus to the rich and powerful, even after he was arrested during the turmoil of the French Revolutionary Wars. He gave sea silk caps to Ferdinand IV, the king of Naples, ordered ladies' gloves for the court of Saint Petersburg, and may have arranged for the shipment of the raw threads to the philosopher and diplomat Wilhelm von Humboldt, brother of Alexander.

The maestro also sent her creations to the powerful: she reportedly wove a sea silk tie for Bill Clinton, a stole for Pope John Paul II and a rosary for Benedict XVI. Yet, unlike Capecelatro, she rejected the commodification of sea silk: 'Byssus cannot be bought or sold, and does not submit to the laws of the market, because it is a collective good: it can only be gifted or received.' She disdained all those who sought to profit from byssus, from the Sardinian medical doctor Giuseppe Basso-Arnoux, who tried to work byssus in mechanical looms ('Only the mind of a man could think up something like that!'), to the weaver Rita del Bene, who proposed the creation of a byssus industry in Taranto in the 1930s. It is said that museums all over the world exhibit Chiara's work, but she never accepted payment, living off donations to her *cassa delle offerte*, an offerings box. 'I would never manage to persuade an Indian chief to sell his headdress. It was not made to be sold,' she told me.

When I visited Chiara's museum, I had spent time in remote caves in Borneo denuded of swiftlet nests, and coffee plantations in Bali stuffed with civet cages. Again and again I was taken aback by the seemingly limitless power of the market, its ability to penetrate every nook, every cranny. After these experiences, Chiara's creation of a sacred space, a bulwark against the market, resonated strongly with me, a pleasant contrast to tales of habitat destruction or cruelty. Every year thousands of tourists come to listen to her, scores of journalists have written about her, and in 2008 the Italian government named her Commander of the Italian Republic, one of Italy's orders of knighthood. Sitting opposite her, I too could not resist her allure. Drawing on her knowledge of the *Pinna nobilis*, passed down through no fewer than thirty generations, she had discovered a way of continuing an ancient tradition while protecting an endangered giant mollusc. 'There is not a single sign on the beach that says that these are a protected species. . . . There's no education in school,' she said. 'So who then protects the *Pinna nobilis*? Me.'

Yet there were, I learned, other voices, which recounted a history of byssus quite different from that of the maestro. For centuries, byssus or sea silk has been the subject of countless confusions, errors, exaggerations and fantasies, a case study for the perils of historical writing. In 1998 the American science historian Daniel McKinley published a detailed study on sea silk. An obsessive and thorough investigator, McKinley pursued and tested the myths, fables and tales that have been told about sea silk over the centuries: Egyptian mummies wrapped in sea silk; Jason's quest for a sea silk fleece; and Henry VIII's tunic supposedly made from the 'beards of bivalves'. McKinley concluded that many of these stories, including a few told by the maestro, were in fact fictions, the result of linguistic misinterpretation or simple invention.

Much of the confusion around the history of byssus can be dated back to a linguistic error in the fifteenth century. In 1476, Theodorus Gaza, a professor of Greek language and literature, mistranslated a simple sentence in Aristotle's *History of Animals* that related to the *Pinna nobilis*. The nature of Gaza's mistake requires more space than I can offer here, but essentially he unwittingly invented a new word to describe the pinna's fibres: *byssos*. This would not have mattered much had it not been for the fact that the word *byssos* already had a set meaning in Ancient Greek (a range of fine vegetable fibres) and appears in numerous ancient sources, from the Rosetta stone to the scriptures. Despite the objections of contemporary naturalists, Gaza's error took root, sea silk became known as byssus, and ever since, writers have confused references to vegetable fibres with sea silk in ancient texts.

Yet there was more than confusion in Chiara's museum – an active and impressive impetus to fictionalise that went beyond mistranslation. As I read sea silk histories, consulted books about Sardinian archaeology, spoke with the Italian coast guard, read up on EU legislation, emailed sea silk experts, and chatted with locals, I found little to confirm the existence of Chiara's millennial thread, one that had started with Princess Berenice and continued up to the present day. Nothing Chiara told me made sense. If I had any doubts about Princess Berenice, Chiara told me, I should read the work of Antonio Taramelli, an Italian archaeologist who had excavated Berenice's tomb on Sant'Antioco in the early decades of the twentieth century. When I finally tracked down a copy of Taramelli's study, I found that he did indeed mention the tomb 'of a woman named *Beronice*', but he'd dated it to the fourth or fifth century, long after the period in which Princess Berenice is said to have lived. If I had questions about her concept of the maestro, Chiara said, I should contact a professor of Hebrew studies at the Sorbonne, but when I did so, I learned that the individual in question had no claim to any affiliation with the institution.

Then there was the question of Chiara's special technique to collect byssus, her repeated dives. In all the reports I had read, no one had actually accompanied her on her special harvests. Did she really dive down to trim byssus beards, wrenching the molluscs to the surface and returning them to their original place? In 1795, Giuseppe Poli, the Italian conchologist, recorded the means by which divers would remove shells. 'Since it cannot be loosened even by repeated blows, (for the sand firmly resists the attempts of the diver, being supported by its own weight and by the super-incumbent water,) in these circumstances he sits down at the bottom of the sea, brushes away with his fingers the earth which encompasses the shell, and then endeavours to pull it up by seizing it with both hands.' Much as I tried to imagine Chiara carrying out this feat, I struggled to imagine her diving down and wrenching these great shells to the surface, clipping and returning them.

As proven by the *Costa Concordia* operation, marine biologists have managed to remove pen shells from their moorings, relocating them without causing long-term harm. In theory, perhaps Chiara could do the same, but this raised another question about the number of beards she would have to trim. In her book *Spirals in Time*, Helen Scales, a marine biologist and shell expert, considers the mechanics of Chiara's technique. She writes that the *full* beards of fifty pen shells would yield only about an ounce of sea silk; given this, Chiara would have to trim the beards of thousands of shells for her annual harvest, previously reported on her website to be six hundred grams (about twenty ounces). Rather than a shaman, she would have to be an athlete to carry out such an industrial operation. Many in Sant'Antioco with whom I spoke thought it simply impossible. Far from a coelacanth, Chiara's history started to resemble P. T. Barnum's mermaid, a creation carefully constructed to please the crowds.

It felt uncomfortable raising doubts about Chiara's personal story in Sant'Antioco, tediously fact-checking her claims. Somehow it seemed that I was missing the point of her museum, interrogating a playful story, a living myth. What did it matter if Chiara wove fictions about a mollusc? I'd studied photographs of the *Pinna nobilis*, read scientific tracts on molluscs and their nervous systems, but it was not until I visited Chiara that the *Pinna nobilis* took on life, transcending its use as flesh or fibre. Many museum curators, storytellers and writers weave narratives, omit facts and tolerate falsehoods to tell a good story, to entertain the crowd or a reader, or to give meaning to objects. Why couldn't Chiara, too? (Which point reminds me to disclose that in this chapter I couldn't resist plundering a selection of Chiara's quotations from the press, mixing them with my own material to give the impression of one seamless encounter.)

I did not seek to answer these questions until I read an article

from 2014 in a Sardinian publication, which reported on an anony-
mous online campaign in Sant'Antioco. It received little publicity, certainly compared with the countless newspaper articles, interviews and
documentaries that featured the maestro. 'For too long,' it began, 'the
truth about byssus becomes little by little rejected, denied, demeaned,
suffocated and frustrated. . . . To such a misleading din we would like
to bring a dignified truth.' The campaign charged that Chiara Vigo
had rewritten the local history of sea silk weaving in Sant'Antioco;
its true master was a craftsman named Italo Diana, who had revived
sea silk harvesting on the island, teaching a group of local women to
work the golden threads. In all of the maestro's talk of Greeks and
Chaldeans, of Aristotle and Aramaic chanting, the story of Italo
Diana had been forgotten or erased.

The byssus that I first described came from Italo Diana's studio in
Sant'Antioco. Some ten years ago, his daughter Emma was
clearing out her father's house on the island after its roof collapsed.
By chance she found two bags in a storage cupboard alongside old
books and cans. 'At first I thought the builders had left garbage behind,' she said. She was tempted to throw the bags out, but just to
make sure, she checked inside and found that they were full of what
looked like tangled dark brown hair, soaked in rainwater. She took
the tangles back with her to her place in Cagliari and left them out
to dry on her balcony. Scorched by the sun, they gradually lightened, morphing from balls of tangled hair to golden locks. 'They
were transformed,' she said, 'as you see them now.'

I'd headed to visit Emma Diana at her small flat after reading
about her father. Spotless and with its dark wooden furniture, her

living room reminded me of many I had visited over the years in Italy, except its walls were covered with curious golden-brown embroideries of different creatures and plants. With a jolt I realised that we were surrounded by sea silk, the fibres reworked to represent the familiar forms of peacocks, birds and flowers.

At the time of Emma's discovery, her father had been dead for some fifty years, but she remembered spending time in his atelier in Sant'Antioco as a young child. He was a great innovator, she said, learning to weave from a young age in a household dominated by women before setting up his own textile atelier in Sant'Antioco in the 1920s. Back then little was known about sea silk, but Italo Diana started experimenting with the thread. He would pay fishermen to retrieve giant pinna shells for him, cut off the beards, and clean and comb them, transforming the brown moss into the golden thread that his daughter would later find in the cupboard. Once cleaned, the beards would be spun and woven to make beautiful objects: tapestries and articles of clothing for the wealthy. It was difficult work, and to help him he employed a group of young women from Sant'Antioco. A photograph from that period shows them sitting outside his atelier with baskets of byssus, fixated on the mossy beards. Anyone unfamiliar with the history of sea silk might presume them seated outside a barbershop, working discarded clumps of hair.

I picked up another object, a miniature child's jacket covered in loops of sea silk, a traditional Sardinian weave known as *pibiones*. Running my hands over the loops, I was charmed by the thought of a child bearing the beards of mussels, this marine anchor converted into elaborate embroidery. The Victorians were fascinated by sea silk, enthusing endlessly over its sheen. Part of the attraction, I thought, lies in the impression of sensory disorientation, the feeling that you are handling human hair or horsehair rather than the fibre of a mollusc. Like many who touch sea silk, I had certain expectations of what marine produce would feel like: the blubber of the whale I had seen in Iceland, the sandpaper skin of sharks, the ground shells of mussels, the flesh of fish. Perhaps this explains why ancient texts refer to 'water sheep' or monsters as the source of byssus; it is easier to believe that fictitious creatures rather than molluscs could produce such fibres.

Italo Diana's objects spread far and wide to exhibitions. One of the largest items of sea silk made in his atelier was a linen tapestry. More than a metre in length, it was covered in embroideries of fantastic creatures and plants, the byssus of tens if not hundreds of giant molluscs. I examined a photograph of it, where it seemed a symmetrical, ordered creation: the patterns on the left and right were perfect reflections of each other, but on closer inspection there was a small section near its centre whose stitching seemed messy, almost chaotic by comparison. Look closely, Emma told me, and you can see the traces of a dedication in its centre, now stitched over – 'W Il Duce' (Evviva [Long Live] Il Duce) – revealing its hidden history.

In the same month of the discovery of the coelacanth in South Africa, December 1938, an Italian cruiser docked at the port of Sant'Antioco and dropped off a special guest: Benito Mussolini, Il Duce. Dressed in full military uniform, Mussolini posed for the cameras of Istituto Luce, Italy's cinematic propaganda organ. He shook hands with various officials while the cruiser fired its cannons in salute. For much of its past, Sant'Antioco had remained on the periphery of history, a berth to such figures as Admiral Nelson, rather than their port of call. But on that December day, the island was placed at the centre of historical events, the subject of countless propaganda reports disseminated all over Italy.

Il Duce had come to Sardinia to open a new fascist town, Carbonia, which lay a short train journey from the island. It would be his third new town in Sardinia, but this one was of particular interest to the Italian dictator: it was built next to an enormous deposit of coal. Sardinia's coal was something of an obsession for him. In 1935 the League of Nations had introduced sanctions against Italy in retaliation for Mussolini's adventure in Abyssinia; he hoped that Sardinia's black treasure would free him from reliance on imports, allowing him the liberty to wage war as he wished. A photograph from 1942 shows Mussolini touching a lump of Sardinia's coal with leather gloves, caressing it as if it were gold, the answer to all his foreign policy problems.

The arrival of Il Duce was a huge event for the inhabitants of Saint'Antioco, then a poor fishing village. In Mussolini's grand scheme, the island was set to become the transit point for Sardinia's coal. Once extracted, the coal would be transported by rail to the island's port and then loaded onto ships. 'There was hope that he'd bring jobs,' recalled Emma. 'At that time there were just fishermen, dockworkers at the port and an elementary school. It was very, very poor.' In anticipation of Mussolini's arrival, a huge pile of coal had been dumped in Sant'Antioco's port. Large letters on the coal spelled out a commitment to hard work, bright white against the dark carbon: 'Duce noi dormiamo con la testa sullo zaino' (Duce, we sleep with our heads on our rucksacks). But there was another, less conspicuous gift for Il Duce. Emma told me that her father and one of his students wove an enormous tapestry made of sea silk with a dedication to the Italian leader. 'He wanted to offer it in the name of the [textile] school,' she said, 'to see if fascism was interested in this kind of work. If it might give subsidies.'

After landing at Sant'Antioco, Il Duce headed to the new town's main square to give a speech. Sant'Antioco's inhabitants followed him, piling into train carriages. Among them was Emma, then no more than seven years old. 'I got on the train without my parents, without anyone,' she recalled. 'Everyone went.' The new fascist town was unlike anything Emma had seen before. With its wide-open spaces, greenery and airy buildings made of local trachyte and limestone, the town drew on the garden city, but it was the logic of the mine that reigned supreme in Mussolini's creation. Each miner's house reflected his status down below; the town's three roads all led to the pits' entrance, feeding the mine with labour. And lest anyone should forget why it was built, Mussolini had named it Carbonia: the land of coal. 'In its name,' he told the inhabitants, 'lies its origin, its duty, its destiny. It will have as its emblem a miner's lantern!'

Sant'Antioco's residents crammed into Carbonia's main square, Piazza Roma, together with thousands of others from all over Sardinia. Flanked by Blackshirts, Il Duce addressed the crowd, his voice veering from staccato to piano. Sardinia was a wild, ancient place: a 'near-deserted land: not a man, not a house, not a road, not a drop of water: [only] solitude and malaria.' But with the arrival of fascist engineers and architects all this had changed: roads had been built, people had arrived and a new town had arisen from the dust. This 'forgotten land' was forgotten no longer! After his speech, Mussolini apparently left as quickly as he had arrived. But Emma told me that the tapestry stayed in Sant'Antioco. The local fascist authority, the *podestà*, had insisted that her father give it on behalf of the community, but he wanted it to represent the school. 'He said no – the community never gave him any support.'

Italo Diana was not the only weaver who tried to find a place for byssus in 1930s fascist Italy. In Taranto, a weaver named Rita del Bene sought to convince the fascist authorities that byssus could make a 'notable and decisive contribution to Italy's economic self-sufficiency'. She planned to industrialise byssus production, raising pen shells and working byssus with mechanical looms. Strong, moth-resistant and insulating, byssus, she said, could be a useful fabric for the new Italy, woven to make aircraft covers, upholstery, a substitute for furs, knitwear and even gas masks. 'I am fully confident,' she wrote, 'that industrialisation will be attainable and that Imperial Italy will find the superb means for its coveted economic self-sufficiency in its own sea.' But no fascist official supported del Bene, and her school closed. Like Italo Diana's gift of sea silk, it came to nothing.

It always surprised me that sea silk, like eiderdown, was proposed as a material for totalitarianism. Beyond its practical limitations, the fibre just didn't strike me as a particularly Fascist material, an effective vessel for an ideology based on strength, national unity and obedience. Admiring a photograph of Italo Diana's tapestry, I wondered whether it was perhaps lucky that Il Duce had not seen it. In his speech he had depicted Sardinia as a desert, a blank slate, whose backwardness and isolation could be overcome only by its black 'treasure'. But the tapestry's ornate patterns told a different story about Sant'Antioco and Sardinia. A celebration of the imagination, craft and mastery, it was a bestiary rather than a propaganda piece. Less a fabric of fascism, sea silk seemed the perfect thread for fiction, a carrier of fantastic stories, creatures and dreams, as Chiara had so well understood.

Rather than sea silk gloves or hats, Il Duce and his entourage preferred to wear black uniforms of orbace, a harsh Sardinian fabric made of sheep wool. Durable and impermeable, it was often worn by soldiers, their bodies protected from the rain by its waterproof matting of fibre. As the membership of the Fascist Party began to swell, more and more people needed the dark fabric of the Blackshirts. Sardinian artisans started weaving orbace for the new fascist market, the material travelling well beyond the confines of their island to mainland Italy. I imagined that the 1930s and 1940s were the boom years for these artisans, but there might have been a cost to the new fascist fashion, a loss belied by the increase in demand; in this new world of blacks and greys perhaps there was little room for Sardinia's other colours, the bright white of linen or the burnished gold of sea silk.

In 1939, Italo Diana found himself entangled in this new politics of weaving: he was drafted into the Italian army, his looms were apparently requisitioned to make orbace, and his school closed for good. It's unlikely that he ever worked with sea silk again. After the war, he took a job teaching textile art at the State Institute of Art in Sassari under the directorship of Filippo Figari, an Italian painter. At some stage he – or someone else – decided to stitch over the reference to Il Duce in his tapestry, masking sea silk's brief flirtation with fascism. (It proved harder to rid orbace of fascist ideology. After Achille Starace, secretary of the National Fascist Party, was shot and strung up next to Mussolini in Piazzale Loreto, a rhyme circulated in Bologna, the Red City: '*Qui giace Starace / vestito d'orbace / dal volto rapace / dall'occhio merdace / vile e mendace / di nulla capace / Requiescat in pace*' [Here lies Starace / dressed in orbace / a rapacious face / with shitty eyes / cowardly and mendacious / capable of nothing / Rest in peace]. Even today the fabric is synonymous with

Fascist Blackshirts; proofed against rain, the dark material fully absorbed Il Duce's ideology.)

Little is now left of the school that produced the Duce's tapestry or its looms. It lies on Via Magenta, a small side street a short distance from Sant'Antioco's centre. One day I walked there from the town's main square, Piazza Umberto, with Claudio Moica, a local poet and writer, who has written about the history of sea silk on the island. An Italo Diana enthusiast, Claudio had interviewed the descendants of weavers and traced written sources to build a picture of the man. 'He was someone who dared,' he told me. 'He experimented with everything. But he was also awkward, silent, introverted.' Claudio reminded me that in the 1920s and 1930s there were not many men who were weaving; at the time this male weaver must have cut a curious figure in a conservative society with clear delineations between the roles of men and women. We stopped outside a nondescript house whose ground floor was a stationery shop, before walking round the back to find a ruin, all rusted ironwork, cracked plaster and broken green shutters. Outside there was a sign warning against unauthorised entry, and a notice from the town council. 'There's little left,' Claudio said, 'but if you look closely, you can still see the taps that carried the water used to work textiles.'

It was striking to see this building, to have followed the golden threads back to this point. By all accounts, this was where byssus was reinitiated in the twentieth century, and yet it was a forgotten place. 'It makes me sad if my grandchildren, great-grandchildren . . . do not know about Italo Diana – do not know about byssus,' Claudio said. 'He was one of the first people who brought Sant'Antioco beyond the confines of an island.' More than the loss of a building or looms, it seemed like a tradition itself had faded. After the school closed, Claudio said, the women went their separate ways, teaching weaving in schools or within their homes; the last surviving weaver from the school, Efisia Murroni, died in 2013. I envisioned a tapestry with a simple design – a single point with separate golden threads branching outwards, but they were cut short, extinguished, as each weaver died.

Standing there, I wondered what the relationship was between this school and Chiara's millennial tradition, the single thread that stretched from the first century until the present day. The maestro hadn't mentioned Italo Diana during our conversation, though she had been scathing about the attempts to commercialise byssus in the 1920s and 1930s in Sant'Antioco and Taranto. 'When someone touches byssus who is not a maestro,' she told me, 'bad things happen.' I asked Claudio what he thought. He said that Chiara's grandmother had been one of Italo Diana's students. I did not know whether to believe him (Claudio had little time for the 'esotericism' of the museum), but later I started reading old interviews with the maestro. Asked in 2000 how she knew how to work byssus, Chiara didn't mention Berenice or a millennial chain connecting grandmother to granddaughter: 'I learned it from my grandmother, Maria Maddalena Mereu, who in turn was a student of Italo Diana.'

I left Claudio and headed north on foot. The *maestrale* wind had dropped and it was sweltering, desert temperatures. Swallows shot through Sant'Antioco's cobbled streets at waist level, banking at intersections. Gradually the houses dropped away, replaced by high hedges of prickly pears, juicy and fat. Nearly delirious in the heat, I grabbed the tip of one of the plants, a huge spiked pad the size of a tennis racquet head, and twisted. It came loose in my hands and I plucked its needles as I walked, sprinkling them on the side of the road and relishing the pad's heaviness in my hands. I enjoyed probing Chiara's story, examining its seams, but by the time I had removed all the spines, I began to feel a light anxiety. For all of Chiara's inventions, falsehoods and exaggerations, I was not sure that I wanted to replace her story with my fragmentary portrait of Italo Diana, this talented weaver and impresario, who dedicated himself to a range of materials in addition to byssus. As a *maestro d'arte*, a teacher of textiles at Sassari, did he care as much about byssus or the pen shells as did the *maestro di bisso*? Why hadn't Italo Diana taught his own daughters how to weave?

While in Iceland and Indonesia, I'd seen first hand how women, often silently and with little compensation, carried out so much of the hard labour required to turn a raw material into a duvet filler or a delicacy. Confined in large processing centres, thousands upon thousands of women clean birds' nests in Indonesia and Malaysia, while Icelandic grandmothers still pick away at down, a painstaking process, in remote farms in the Westfjords. Chiara's sea silk narrative was so different, an exception in which a woman was in full control of an object's pipeline. As a man, I didn't want to be the person to question it, so when I returned to speak with her in her museum a few days later, I still hoped that these two threads, that of Berenice and that of Italo Diana, might run in parallel through time.

This time I wanted to get Chiara alone, far away from the tourists and the demands of her daily performance. During our first meeting, I had sensed a slight hint of mischief, as if she knew that the duration of her oral tradition, the number of deep dives, the standard of her Aramaic wailing, were slightly suspect, a ruse to entertain visitors like me. I hoped that alone, or without outsiders, she might reveal more of this side, but I was to be disappointed. It was late in the evening and, though the tourists had long gone, her room was full of her students, many of whom came from outside

Italy. As luck would have it, she was already on the subject of Italo Diana, striding back and forth for dramatic effect.

'Italo Diana did not know how to weave!' she said, clapping her hands to the syllables. 'Show me what pieces Italo Diana made! Where are the pieces that Italo Diana made? The Duce's tapestry was made by Assunta Cabras! The jacket and cap were made by Caterina Dessi! The tie was made by Efisia Murroni for her husband!' The atmosphere, I noticed, was different from my earlier visits. The charm, the smiles, had gone, replaced by anger, and my own presence, my questions about Italo Diana, only made things worse. Her voice was different, too; it was deeper, the voice of a smoker. 'Italo Diana was not a maestro!' Chiara cried. 'He was a swindler who did not even know how to weave. He took advantage of women who did know.'

The history of textiles is largely one of the exploitation of female labour, often directed by men. Chiara's Italo Diana was a continuation of this narrative, another man who exploited women. His was not a school, Chiara said, but a place of commerce where women were victims rather than students. 'I have looked on land and sea and never found any evidence of a school,' she told me. He was just like all the other sea silk entrepreneurs of the twentieth century, a charlatan looking to make money from the sacred fibres. 'Be careful about what they call documents [about Italo Diana]! They are writings that are invented.'

It was fun turning Chiara's tapestry over to view the fibres of which it was constructed, unpicking her strands. But here I saw a darker side to her work, an impetus to destroy that ran parallel to her narrative creations. While she wove new historical patterns, creating stories about a princess of byssus and a millennial tradition, she started pulling the threads of local history, undoing the patterns of other weavers.

'Why is there a museum of byssus?' she went on. 'Who created it? How do we know about byssus in Sant'Antioco? Italo Diana? No.' It was here that Chiara had a point. The world knew about byssus because of Chiara Vigo. She had made sense of its complex history, brought to it life, refashioning it to carry her own name. 'This exists,' she said, taking some byssus in her hand, 'because I exist.'

I felt oddly afraid – not of Chiara but of the effect of her words on those around her. I knew her words to be false, that her expert witness was not a professor at the Sorbonne, that Taramelli had not discovered the tomb of Princess Berenice, but I felt isolated in her museum. As she trashed Italo Diana, weaving a new historical reality, her students nodded at her words or laughed at my questions. 'You need at least two years to carry out a detailed study!' scolded Giuseppe. 'Sometimes we die, like Italo Diana,' said another student, laughing. 'So anyone can say anything about us!'

The Italian word *tessuto* (fabric) shares its root with the word *testo* (text). Later, as I reflected on the museum, it struck me how much Chiara resembled a high priestess interpreting the meaning of a textile – byssus – as if it were a sacred text or arrangement of entrails. Less important was the meaning of the text or intestines than the very concept of a secret, hidden information that must be decoded, processed, interpreted, disseminated, but never revealed by the elect, the maestro. 'I am and must make history,' she told me. 'I must pass on stories. After me, Maddalena . . . And so it continues. To continue a story that dates back to the beginning of time.'

I left the museum and headed along Via Eleonora d'Arborea, the street that headed along towards the town's piazza. Walking alone in the dark, I felt strangely empty. After so many hours in Chiara's museum, I had learned little about the creature that actually produced sea silk; within its walls the wild mollusc was merely a silent prop, a dead object onto which her own dreams could be projected. Had the *Pinna nobilis* been absent from Sant'Antioco's lagoons, perhaps she would simply have found another object—a sceptre, sacred text or lump of coal – with which to amass power.

What explained Chiara's hatred for Italo Diana? Why was it necessary to remove the skill that had defined his life, his ability to weave? In all that I had read, including testimonies of the descendants of weavers, no one had ever described him in the terms used by Chiara, nor was there any proof that he did not know how to weave sea silk. Yes, his presence dampened Chiara's fantasy, but the foundations of her museum still stood, while those of his school were ignored or forgotten. Insert Italo Diana into Chiara's story, and she remained an accomplished weaver with an engaging ability to tell stories. In any case, all of Italo Diana's students who learned how to work byssus were dead; she reportedly remained the last person who worked byssus. What else did Chiara have to fear?

I turned back to the voluminous output of interviews, the textual record of her oral tradition that went back to the 1990s. In Chiara's first interview, from 1994 in an Italian popular science magazine, Italo Diana is described as 'the true and real "discoverer" of byssus in this century', who passed his knowledge on to Chiara's grandmother. Yet after 2000 there is a clear shift: Italo Diana's name no longer appears in articles on byssus. Dropped from the narrative, he is replaced by stories about Berenice and discourses on the hereditary nature of the maestro. What had happened around 2000?

The island of Sant'Antioco takes its name from a Christian physician, Antiochus, who was said to have been born in Mauritania, then part of the Roman Empire, in the first century AD. The story goes that his abilities as a healer eventually attracted the attention of the Roman emperor Hadrian. Summoned to Hadrian's presence, Antiochus prayed to God, upset the emperor, underwent various hideous tortures and was exiled to the island off the coast of Sardinia. There he continued his good works until his eventual arrest and execution by the Roman governor of Cagliari.

The saint's remains are held in the church that bears his name, a short walk from Chiara's museum. I admired his bones displayed in a glass case while I waited to meet Assuntina Pes, another weaver who knew how to work sea silk. I caught sight of her as she laid a rose petal by a statue of Sant'Antioco. Turning around, she offered me one, too. We drove to her house on the outskirts of Sant'Antioco, picking up her sister, Giuseppina, another sea silk weaver, who lived just next door. Even after both of them got married, they insisted on being close to each other. 'Better next to my sister than anyone else,' Giuseppina said, laughing. 'Otherwise you'd argue.'

Had it not been for their hair, it would have been hard to tell the two sisters apart. Giuseppina had long locks with red streaks, while Assuntina had darker curls whose colour reminded me of byssus. Whenever one started talking, the other would finish off her sentence, and later, when I listened to our recorded conversation, it was easy to think of them as a single person. After we pulled up to Assuntina's house, they led me into a cramped studio dominated by a traditional loom, its walls adorned with tapestries whose patterns evoked the work of Italo Diana.

The two explained that they had first learned to weave in a school run by Chiara's grandmother Maria Maddalena Mereu. In the 1980s, after finishing the course, the sisters opened their own cooperative, Sant'Antioco Martire, weaving with cotton and wool and teaching other women. Then, at the end of the 1990s, they got to know Efisia

Murroni, one of Italo Diana's students, who taught them how to work sea silk. 'She didn't have a school, she didn't have anything,' Giuseppina recalled. 'As we could already work with wool and linen, she just showed us the various steps [to work with byssus], how to hold the spindle, how to clean it, the various stages. Until [one day] she was happy and told us: You can now walk on your own.'

Over the past couple of decades the sisters have produced a number of works, the first appearing in public in 2000. Instead of working with fresh byssus, the sisters told me, they used old material from Italo Diana's school, the byssus that had been discovered by his daughter in the ruined house, or fibres gifted by their old teacher Efisia. The quantity was limited, but Giuseppina said that made it more special. 'It's a beautiful feeling knowing that it came to the school and that there were other women who worked it.' Admiring one of their tapestries, I was struck by how different their conception of byssus was from that expressed in Chiara's museum. From Italo Diana's studio, many threads had expanded outwards, the knowledge of sea silk diffused. It had passed from women to other women without rites or ceremony, the necessity of shared blood.

The sisters laid out about six clumps of byssus on a low wooden table covered with white linen cloth. The first resembled a clump of seaweed tangled with countless white shells. As I moved around the table, I saw that the fibres gradually straightened, the shells disappearing to reveal dark brown locks of fibre, each strand about the thickness of a human hair. 'It takes a lot of time,' Assuntina said, the excitement in her voice palpable. 'You must clean it very gently, delicately, so the fibres do not break.' I asked them whether they could demonstrate the various steps to clean, spin and weave, but Giuseppina said that they had no byssus to spare. The *Pinna nobilis* became a protected species in 1992; no one is permitted to touch a shell, let

alone harvest its threads. 'There is no byssus – you cannot collect it,'
Giuseppina said. 'Otherwise we'd go to prison.'

n the Middle Ages, many fictions were woven around crustaceans
and other sea creatures. It was commonly said, for instance, that
birds grew on trees or hatched from barnacles, sprouting forth as
fruit. Around 1356, Sir John Mandeville (or someone identified as
such) wrote of 'trees which bore a fruit that became birds that could
fly, men call them bernakes, and there is good meat on them. Those
that fall in the water live and fly away, and those that fall on dry land
die.' As late as 1677, Sir Robert Moray, a member of the Royal Society, dissected barnacle shells that he found on a fir tree in Scotland
and reported that he discovered 'nothing wanting, as to the external
parts, for making up a perfect Sea-Fowl': 'the little Bill like that of a
Goose, the Eyes marked, the Head, Neck, Breast, Wings, Tail and
Feet formed, the Feathers every where perfectly shap'd, and black-
ish colour'd; and the Feet like those of other Water foul, to my best
remembrance.'

The proliferation of these stories speaks of the absence of geese
in the lives of writers; few had perhaps seen geese nest and lay eggs.
But the stories may also reflect certain vested interests, a deliberate
attempt to disguise the truth. If geese sprouted from wood, then they
could not be classified as meat and could be eaten during Lent.
'The "monks of old," and "the barefooted friars," as well as the laity,'
suggested Henry Lee, an English naturalist, 'may not have been
unwilling to sustain the fiction in order that they might conserve the
privilege of having on their tables during the long fast of Lent so
agreeable and succulent a "vegetable" or "fish" as a Barnacle Goose.'

Reading these old stories about sea creatures, I began to wonder whether Chiara's stories did not serve a specific purpose. If it was revealed that knowledge of byssus in Sant'Antioco originated in Italo Diana's school, then no single individual could claim the exclusive right to harvest, clean, weave and tell stories about its threads. The concept of the maestro and its system of rights would unravel; Chiara would be one woman among several and – far from heretics or impostors – the Pes sisters would be legitimate heirs to a local tradition. Before me the landscape of Sant'Antioco appeared anew, clearer, but immeasurably darker. The destruction of a dead weaver, the story about Berenice, the sacred oath, and the tradition of the maestro: all these stories functioned as a means of controlling other weavers, all of them women. Strip the prayers, rites and wailing, and Chiara appeared to me as little more than a common market figure – the monopolist – in the garb of an ancient shaman. Indeed, later I learned that in 2013 Chiara had threatened to take legal action against the Pes sisters if they ever worked with byssus, in order to protect her hereditary rights.

I left the Pes sisters, and one of their friends gave me a lift back to Piazza Umberto in her car. As we passed the Byssus Museum, I ducked my head, worried that Chiara might spot me on my way back from visiting the sisters. 'Ma cosa fai?' the friend asked, laughing. 'Hiding,' I said. Like gravity, Chiara's power was easy to sense, but hard to explain. I felt its force in the silences of the Pes sisters, the reluctance of the volunteers in the island's ethnographic museum to talk about the history of byssus beyond the 1930s, the absence of Italo Diana's name in Sardinian newspapers, Chiara's disregard for national legislation on protected species, the lack of byssus experts in Sant'Antioco, the silences of marine biologists on the subject of Chiara's special technique and my instinctive attempt to hide in the car.

Few on the island were prepared to challenge her authority. One person who did was the local woman who dropped me off in the car, Antonella Senis. She was in her sixties, with dark hair and a calm, quiet voice. Unlike the Pes sisters, she was outspoken about what was happening on her island, and was unafraid and deeply angry. In Chiara's tapestry – one replete with images of her family, deep dives and secrets – she did not see her or her island's past reflected in its florid patterns. 'It's not our history,' she said. 'We do not want our grandchildren to inherit something that is not true.'

In her spare time, Antonella worked with a local activist group, campaigning against pollution in the Sulcis region. She spoke of the Byssus Museum as another great polluter, a dirty factory spewing noxious fictions into the landscape. 'It's hard to explain,' she said, clutching her chest as if to calm an ache. 'I have this feeling that something that is ours, something that is mine, is being taken away from me.' It is normally the state or the press that challenges monopolies: muckraking journalists, regulatory bodies, competition commissions. But in Sardinia most of the press and state officials had tacitly accepted the rules of byssus and Chiara's universe. Newspapers carried countless stories about the 'last maestro of byssus'; even certain academic publications continued to refer to Chiara as the last master.

In the absence of a competition commission for weavers, Antonella told me that she tried to break the sea silk monopoly alone. In 2014 she set up a Facebook group and started posting research about byssus from peer-reviewed journals and interviews with experts and descendants of weavers. She called the group 'The True History of Byssus in Sant'Antioco'. Looking through Antonella's postings, I was struck by this collection of heretical objects: the photographs of Italo Diana and his students, their sea silk objects, biographical information about other weavers, and a letter from the coast guard stating that the *Pinna nobilis* was protected: no one had the right to even touch it. Taken as a whole, her pages presented more than a different history of a trade. It was a different history of an island – one with many branches rather than the single thread stretching back to antiquity. 'It's a drop of truth in a sea of falsehood,' she said.

Antonella's efforts to return Chiara's autobiography to the fiction section of the bookshop did not have much success. As she placed primary sources online, nothing much happened; the tourists kept on piling into the museum, the donations rolled in, and Chiara's reinvention of the island – mysterious and ineffable – continued to proliferate, virus-like, in glossy brochures and the travel sections of newspapers. 'We live in the land of Pinocchio,' Claudio Moica told me. 'People prefer the myth.' But there was one figure in Sant'An-tioco who had the power to unseat the sea silk priestess: the town's mayor, Mario Corongiu. Under a ten-year-old agreement, Sant'An-tioco's council had leased the museum's building to Chiara at no charge; in theory the mayor could end it when he wanted. At the end of 2015 he did just that. Two days before Christmas, Chiara received a letter from the council saying that she had to vacate the museum. 'They ruined my Christmas,' she told the press. 'I feel betrayed, the museum cannot be moved from here, if they send me away I will go abroad where my works are already appreciated.'

In his letter to Chiara, the mayor justified his decision by citing health and safety concerns about the building's electrical wiring, but his move was widely interpreted as a pretext to rid Sant'Antioco of its turbulent priestess. Whatever his motivation, the letter was greeted ecstatically by those who disliked the museum's style of storytelling. Any celebration, however, proved short-lived. A master strategist, Chiara promptly enlisted the support of a sizeable group of influential backers: politicians, celebrities, local fans and artists.

There were offers to carry out maintenance work in the museum for free, mayors throughout the Sulcis region pledged to host the maestro, a vigil was arranged outside the museum, and an online group was formed in its defence: 'Nessuno tocchi il Museo del Bisso' (Hands off the Byssus Museum). 'When a maestro is attacked,' Chiara told me, 'she will defend herself!'

Chiara's most prominent supporter was the Italian actress Maria Grazia Cucinotta, best known for her roles in *The Postman* and *The World Is Not Enough*. Shortly after Chiara received the mayor's letter, Cucinotta set up an online campaign to save the museum and 'the ancient knowledge held since the beginning of time by her family'. The unlikely trinity – mollusc, maestro, and Maria Grazia – proved electrifying; before long, the campaign had close to twenty thousand signatories (clicks), almost double Sant'Antioco's declining population.

Weaving has often been associated with acts of defiance. During the American Revolution, George Washington encouraged his troops to wear homespun uniforms rather than clothing produced by the British; Gandhi also encouraged the weaving of homespun or khaki in India to end dependence on British cloth. As I followed the conflict, I marvelled at how Chiara became something of a resistance hero pitted against the new colonising force of our age, the market. Celebrities, politicians and journalists flocked to her, the last authentic bastion against rapacious commercialism. 'In a universe almost totally dominated by the market and iron economic laws,' wrote Cucinotta, '[Chiara's refusal to sell] makes her a stranger in her own land and in a way exiles her from her own community.'

Like Chiara's supporters, I felt the attraction of this interpretation, but her refusal to commodify byssus seemed to mask another act of commodification: not of a physical object – byssus – but of its history and that of the island where it was harvested. Like any physical product, Sant'Antioco's past could be tweaked, altered, modified to suit the needs of the customer and reap commercial benefits. 'June, July, August, it's always full,' Chiara's husband told me. 'She's the one who brings a load of people here. They go to restaurants, hotels.' Half church, half factory, the museum was one of the island's star attractions, a marriage of mysticism and the market.

In untangling the insults, shouting and threats of litigation in Sant'Antioco, it seemed that this brute reality was at the root of the uproar in favour of the museum. While some appeared to sincerely believe Chiara's stories, behind the chanting, pledges of allegiance and vigils lay a simple calculation. Strip Chiara from the story of byssus, and would any tourists even bother visiting a small community museum? Would anyone even care about its fibres? 'Byssus today is Chiara,' wrote Fabrizio Steri, a local businessman and the founder of the online defence group, 'and without Chiara, no one gives a shit.'

It was a conflict not about past events – names, dates, and facts – but about the nature of history itself: whether it was sacred and inviolable or malleable like any physical product. 'Look at the saint who gave his name to this island, Saint Antiochus,' Fabrizio told me. 'That this man existed there is no doubt, but can we really say that everything that is said about him is true? That he was tortured by Hadrian, placed in a cauldron of burning tar, exiled, and shipwrecked on this island? . . . We live in a world in which everything we derive from the past has inevitably been coloured, passed down.'

Despite the pledge of its inhabitants in 1938 displayed on a pile of coal –'Duce, we sleep with our heads on our rucksacks' – Sant'Antioco never became the heart of an Italian coal industry. Its port is now an abandoned no-man's-land, an invisible space. 'It was once the most important port in Sardinia,' said Chiara's husband. 'There was trade – magnesium, grain, salt, minerals. It was always full of ships. . . . Now it's collapsed.' In his 1938 speech, Il Duce had praised the Sulcis coal as a treasure, but the reality was somewhat different: Sardinia's coal was laden with sulphur, a quality that made it extremely costly to process, highly polluting and uncompetitive. In 1964, less than three decades after it was opened, the mine of Serbariu was closed. No one wanted Sardinian coal, not even the Sardinians.

For the inhabitants of Mussolini's new town, Carbonia, the decline of its principal mine proved disastrous. With the mine closed, there was little work; many of Carbonia's men, themselves recent migrants, simply emigrated abroad to find work, leaving their families behind. 'The few people remaining', stated the Sardinian historian Manlio Brigaglia in 1969, 'seem as if they have just survived a sudden cataclysm, which in reality never happened, because the illness from which Carbonia suffers is not from today or yesterday, but was born with the city.'

After the mine's closing, the region was offered some hope by the Italian government's so-called Plan for Rebirth, the development project to build factories and refineries throughout Sardinia. A massive industrial complex was built next to Portoscuso, a fishing village across the bay from Sant'Antioco, to employ local miners. After the plant's privatisation in the 1990s, it struggled to compete in the international market; the region was encouraged to turn to tourism.

Contemplating this history, it is understandable why fantasies about byssus have their attractions. Chiara's stories seemed to work in Sant'Antioco, acting as a narrative fix in a region that produced aluminium that was too expensive and coal with too much sulphur. Turn a thread of byssus into a sacred millennial fibre and Sant'Antioco might recapture its former glory as a pilgrimage site, the island's visitors leaving with bits of byssus rather than relics from a Mauritanian saint.

Unlike narratives about coal or heavy industry, there seemed no price to pay for fantasies about a mollusc: no failing factories, botched mines or polluting refineries dreamed up by Mussolini or the Italian government in Rome. But I also saw that in Chiara's alchemical conversion – the reconstruction of an island's identity for a foreign appetite – there lay historical loss, upheaval and conflict. As the logic of the museum spread, its confines dissolved by the media, the island was slowly becoming detached, uprooted from its own history like a shell from an anchor of byssus.

left the town of Sant'Antioco and drove off towards Calasetta, a fishing village at the far end of the island. It was there, I had been told, that I would find outcrops of pen shells in the lagoon's shallows. Sant'Antioco's lagoon is now a haven for the *Pinna nobilis*. Under the protection of legislation, the numbers of the shells have swelled, forming small nurseries that filter its clear waters. Some believe that the name Calasetta originally meant 'Bay of Silk' in Sardinian, a fitting name for the place to which I was heading.

I wanted to feel changed by my encounter with this giant shell – for it to take its rightful place in its own history, leaving behind the stories, myths, and conflicts with which it had become entangled. I'd read about the beds of mussels found in cold seeps, deep trenches that lie kilometres underwater. Within the animals' gill tissue, biologists have found special chemosynthetic bacteria that fix carbon and manufacture sugars, allowing the mussels to live without sunlight; a mollusc held the potential to change our own concept of life itself.

Just before the village, I turned off onto a dirt track that led to a small beach of bright white sand. Beyond me, a turquoise pool stretched out towards the Sardinian mainland and the chimney of Portovesme's power plant, the tallest structure on the island. The wind had dropped and it was silent, apart from a family unloading a jet ski into the calm waters. Looking around, I found it hard to conceive that there could be a field of pen shells under this clear water: the land was so barren, so devoid of vegetation, apart from rows of prickly pears.

After the jet ski launched off, I put my fins and mask on and waddled backwards into the water. I swam out into the shallow lagoon, its warmth cooling my sunburn. Accustomed to the muddy water of ponds or the mistiness of brine, I was overwhelmed by the clarity of the bay's warm water and the detail that it revealed. Below me lay fields of posidonia, dark green seagrass, and small silvery fish, an underwater forest.

I swam on beyond the grass and found a miniature clearing; it was here that I discovered what I was looking for: a juvenile *Pinna nobilis*, rising upwards from the white sand. Each half of its shell was the size and shape of a hand fan, indented with deep grooves of calcium. Its features were clear-cut, undisguised by large barnacles or the stubble of weeds. Raising my head, I looked beyond the shell and saw countless others, respectfully spaced half a metre from one another. I was floating above a nursery full of saplings, which were quietly filtering the pristine water for its nutrients.

I followed the outcrops of small shells and saw that they gradually grew. Soon I was admiring shells the size of tombstones, covered in great ancient weeds and gnarled with barnacles. From where I was, suspended above these monsters, they seemed inanimate as rocks or half conscious as oaks, dead things surrounded by life: weeds, pea crabs, fish. But as I swam down to them to peer into their mouths, examining the fleshy, frilly space between their shell halves, they would snap shut, locking to form a single seamless whole. It was an act of defence, but I also felt it as a silent rebuke: I am alive.

The dull roar of a jet ski's motor interrupted my reverie. Looking up, I saw that another rider had launched off the beach. Nervous about its motor, I quickly swam back to shore, my chest tickled by the blades of seagrass that formed rafts on the water's surface. I pulled my fins off and admired the man merged with machine, churning and boiling the shallows.

Watching the jet skier, I was struck by the vulnerability of these shells. Millions of years ago they had evolved without consideration for human hands, fabric or machines. Immobile and conspicuous, they could be knocked over by a motor's propellers, plucked with ease by fishermen or caught up in nets. Their only defence in this altered environment was their quietness, their silent rhythm filtering the water as motors roared overhead.

The jet skier returned to the shore, parked the ski and lay down on a deck chair, soaking up the sun. Curious about his views on the shells, I went over to him and asked him about them. He knew exactly where they were, he said, and especially avoided them. As he seemed knowledgeable, I couldn't resist asking him a question that still bothered me. No – he laughed – he did not know how the maestro got her byssus.

VICUÑA FIBRE

n October 2016 the maestro was, to quote the Italian press, finally *'sfrattata'* (evicted); the Byssus Museum was closed for good and the portrait of Chiara was removed. The spell, it seemed, had been broken. In the following months other women in Sant'Antioco, the descendants of the weavers from Italo Diana's school, continued to exchange their stories, perhaps emboldened by the recent developments. Most intriguingly of all, one of Chiara's former students, an accomplished weaver named Arianna Pintus, took the heretical step of working byssus in her own atelier, defying the tenets of the maestro to weave a series of beautiful tapestries. Constrained by EU legislation on protected species, she told me that she sourced her byssus from a different species, *Atrina pectinata*, which is fished and farmed throughout the Pacific.

I observed these developments in the byssus saga with a degree of detachment. While I was excited by the emergence of different voices, ones that turned the monotone of sea silk into a polyphony, I knew that Chiara's story, refined over decades of performance, was just too powerful to be replaced by a history of facts and fragments. One could only hope that in a hundred years' time, future historians, scraping old servers, might find the name of Italo Diana and his school.

In Sant'Antioco the monopoly on storytelling was secure as long as there existed a monopoly on the byssus harvest. Reliant on old byssus from Italo Diana's studio or sources outside the Mediterranean, the silent weavers of Sant'Antioco could never produce works as Chiara did, or gift them so profusely. I did wonder what would happen if the harvest of and trade in byssus were legalised. Could there be a legal, rational and equitable harvest of byssus, one that was not beholden to a single bloodline? I thought back to the attempts to create a rational byssus industry, the experiments of Attilio Cerruti, the Italian biologist who in the 1930s planted pen shells as if they were seeds. Despite his conclusion of prohibitive costs, he had found that pinna aquaculture was in fact possible.

Yet I was worried, too, by the possibility of pinna aquaculture at a time when the shells were still endangered. In 2017, Spanish biologists raised the alarm about a mass mortality event of *Pinna nobilis* in the Spanish Mediterranean, believed to be caused by a parasite. Sacred stories perform a useful function, even if they are based on invention. There is no evidence that Saint Cuthbert ever tamed eiders, but no doubt his mythical relationship with them spurred efforts to protect them. I suspected that Chiara's own stories performed a similar function. Unlike mussels, clams or scallops, sea silk would surely have a market value that might provoke rash speculation or ecological ransacking, one that would rival the emptying of the caves in Borneo. Intrigued, I began to search for a prized object through which I might explore the consequences of legalisation. The more I read, the more I found that the object that most shed light on byssus was not mussels, scallops or clams – whose trade has long been legal – but the fibre of an animal separated from molluscs by some eight hundred million years of evolution.

n 1958 the United States was gripped by a political scandal involving President Eisenhower's chief of staff, Sherman Adams, and a New England textile magnate named Bernard Goldfine. Several years earlier the Federal Trade Commission (FTC) had accused Goldfine's companies of mislabelling products, and he turned to Adams, an old friend, for assistance. Adams duly obliged, which may not have been a problem, except that Goldfine was a prolific gift giver. During packed sessions of the House Special Subcommittee on Legislative Oversight, Adams disclosed that Goldfine had paid for his hotel bills and given him various objects: an 'oriental rug' and an overcoat of vicuña fibre.

Goldfine didn't seem to grasp the problem of giving a vicuña coat to a politician or official. Indeed, it was disclosed during hearings that he had given vicuña cloth to the governors of forty-eight states. 'I don't know any other way to show my appreciation,' he later said. 'I'll continue giving until the end of time.' The press, Republicans and the House Special Subcommittee saw it differently. Adams eventually resigned over 'the vicuña coat affair' and retreated to Lincoln, New Hampshire, where he ran a ski resort, Loon Mountain. 'As I look back now on that whole unhappy episode, these mistakes in judgment are plain enough,' he wrote in his 1961 memoir. Goldfine, for his part, became known as the 'vicuña man'.

Perhaps one could forgive Adams for making such a mistake. Run your hand down a vicuña coat and thereafter the world will feel harsh; lambswool will feel like a Brillo pad, byssus like wire. The coat of a wild camelid, vicuña fibre is often called wool, but its diameter of 12.5 microns is closer to silk than a sheep's coat. By the time the Goldfine scandal broke, vicuña was one of the most prized materials in America, associated with the glamour and social whirl of Hollywood movie stars and musicians; Marlene Dietrich, Groucho Marx, Nat King Cole and Greta Garbo all wore vicuña coats. 'When a mink wants to give a mink a nice coat, he gives it a vicuña,' a Philadelphia tailor told the *Inquirer* in 1958.

There is something strikingly elegant about the vicuña and its curious mane of wool drooping down from its chest. The animals' doelike eyes, their agility and their herding behaviour have all elicited comparisons with antelopes or deer. Yet if one looks closely at them, examines their long necks and padded two-toed feet, they almost resemble miniature humpless camels. They are in fact cousins of the two-humped camel of the Gobi Desert, members of the camel family (Camelidae) that originated and evolved in western North America before expanding across land bridges into Asia and South America. And yet this kinship with the camel must be distant: vicuñas seem more delicate than those robust desert beasts of burden, their necks too long, their branch-like legs too slim. Remove their fibre and, I imagined, they might entirely disappear, merging with the grasslands that they inhabit.

The fineness of the vicuña's fibre is a response not to proximity to the Arctic Circle – as is true for eiderdown – but to height. Creatures of altitude, vicuñas roam in the Andean puna, where the air is so thin that it seems to almost suck warmth from your body, chilling your marrow. When the temperature drops at night, vicuñas lie down on their bellies, their long bibs of fibre forming elegant skirts around their breasts, at once a blanket and a windbreak. Come sunrise, the fibre turns into a parasol, protecting its owners from the sun's rays, which beat down mercilessly on the puna; to counteract its insulating properties, the animals perspire through their bare bellies.

The Incas revered vicuñas, believing them to be sacred creatures. Particularly prized was the animal's fibre, among the finest in the world; only the Incan emperor himself – along with certain priests – could wear clothing made from vicuña. Under the Incas, the vicuñas flourished, but when the Spaniards arrived in the sixteenth century, they showed little respect for native laws or traditions, slaughtering the vicuña for its meat, hide and wool. According to one later report, eighty thousand animals per year were killed in Peru and northern Chile, a massacre evoking that of the buffalo in nineteenth-century North America. Garcilaso de la Vega, the illegitimate son of a Spanish aristocrat and an Incan princess, wrote that 'such havock hath been made by the Guns which the *Spaniards* use, that there is scarce a . . . *Vicuna* to be found, but what are affrighted into the Mountains, and inaccessible places, where no path or way can be made.'

In all the columns, interviews, broadcasts and articles dedicated to Sherman Adams's downfall, we learn little about the fibre of which the coat is made. Unlike the man who wore it, the vicuña coat is merely a prop, a symbol of luxury and excess. Had it been made from mink fur, itself the object of an earlier political scandal in 1951, Adams still would have lost his job, left Washington and spent his days in a ski resort in New Hampshire. Yet Adams's predilection for vicuña did matter. When he accepted Goldfine's gift, he unwittingly entered into another story, one far removed from factory farms, the FTC, House subcommittees or the White House. By the 1950s most mink coats came from farmed animals, raised in battery cages like chickens, ducks or civets, but Adams's coat had once belonged to a wild animal, one that was slowly being driven to extinction.

It's often the case that materials in the natural world that confer on their owners an evolutionary advantage are converted by our own species into dead weights, burdens that hasten their destruction and even extinction. An elephant's tusk, so prized for its density and beauty, has acted as a magnet for poachers; the great height of a Bornean ironwood tree makes it suited for timber; the size, beauty and translucence of the *Pinna nobilis* has made it popular as a lampshade. Nowhere was this principle clearer than in the case of the vicuña and its fibre. By 1967 vicuñas numbered some ten thousand, a fraction of the two million thought to exist at the time of the Incas. 'In the United States,' lamented Felipe Benavides, a Peruvian conservationist, to *The New York Times*, 'many people know the buffalo only through its likeness on the nickel. I wonder whether we will know the vicuña only through our 1-sol piece.'

When Benavides spoke of the vicuña or any endangered animal, he choked with emotion. Unlike political journalists or House investigators, he read Adams's story not as a parable of political corruption or a decline in ethics in Washington, but as another, far greater moral failing: that the demand for vicuña fibre in the United States and Europe, particularly Britain, was driving the animal to extinction. Since the 1930s he had investigated the trade in vicuña fibre, tracing the appearance of the material on London's Regent Street and in Savile Row tailor shops, export ledgers and press articles. Enraged, he wrote to newspapers, publicly shamed fibre importers and lobbied on behalf of Peru's endangered animals in an attempt to make vicuña fibre as unacceptable as cocaine. 'The persecuted majority', he once told a journalist, 'has no voice or vote, no syndicate or unions anywhere. Someone has to speak for them and to protect them from man!'

In the 1970s, *The New Yorker* described Benavides as a 'tall elegant man of fifty-eight with graying hair, a small moustache, and lively eyebrows, and with dark eyes and a high-bridged nose that seem to have come from a portrait in the Prado.' The nephew of a former Peruvian president, he spent the first twenty years of his working life on the diplomatic circuit in Stockholm, New York and London. He appeared destined for a major diplomatic post, but in 1954 he abruptly quit the service and returned home to Lima. After so many years abroad, Benavides was startled by the scale of environmental destruction that he found in Peru. He recalled: 'I went around my country and was shocked by what I saw. I saw people shooting condors and killing sea turtles. I saw wounded sea lions crying like babies, condors left to rot in the sand. Horrible! Horrible! From the shore, I went to the Andes. I could find no vicuña.'

In the 1950s, Peru had no conservation movement. 'What is going on?' asked Benavides in 1954. 'I found that there was not one conservationist in the country. Not *one*! I had to do something.' In 1973 he helped set up Manú National Park, then Latin America's largest nature reserve, and the Paracas National Reserve on the coast. He also lobbied on behalf of a safe home for the vicuña, the Pampas Galeras National Reserve in the southern Andes, founded in 1967. Protected within an area of 5,800 hectares, the vicuña flourished in the park's early years; by 1977, the Smithsonian declared that the vicuña had been saved from extinction.

A photograph of Benavides in his later years shows him nuzzling a vicuña, his large wiry moustache brushing against the vicuña's chin as if it were his grandchild. More than boost vicuña numbers in the park, Benavides changed the very idea of these creatures and their fibre, this fabric that sheikhs, film stars, and at least one presidential advisor were once happy to drape around themselves with casual disregard for its origins. Much like Chiara's vision of sea silk, vicuña fibre entered public consciousness as a sacred fibre, one that could not be bought or sold, at least not without the risk of a three-to-five-year prison sentence. 'Vicuña wool,' Benavides declared at the time, 'looks best of all on vicuñas.'

The story of Benavides and the vicuña resembled a fairy tale, but it soon began to unravel, turning into a nightmare. In the mid-1970s there was a bad drought in Pampas Galeras, which, combined with overgrazing, led to increased adult mortality and reduced pregnancy rates. Fearing a population collapse, Antonio Brack Egg, a conservation official in the Peruvian government, proposed the culling of excess males. 'The cheapest and most effective way is to kill with a Remington caliber .222,' he explained in 1979. A practical conservationist, Brack planned to sell the meat harvested from excess males to benefit the local community; if Benavides was sentimental about the vicuña, then Brack spoke like a butcher. 'The meat is lean, high in protein and of excellent quality for consumption, jerky, drying, and to make stuffing,' he wrote.

In the 1930s New Zealand began culling its red deer population, which was spiralling out of control. Farmed or hunted for their flesh and skins, the deer came to form the basis of a large industry estimated in 2010 to be worth NZ$306 million. Just like the New Zealanders with their deer, Brack wanted to turn the vicuña into a valuable economic resource. 'It would be possible', he argued in 1987, 'to reach a population of three million vicuñas, the rational management of which could produce 225 metric tonnes of fibre a year, about six hundred thousand skins, and some twelve thousand metric tonnes of meat.'

Although red deer and vicuñas share some physical traits as well as social structures, the history of these two animals could not be more different. An invasive species, the deer was introduced to New Zealand in the nineteenth century from deer parks in Britain to turn the islands' 'waste country' into 'a very paradise for the deer-stalker'. The vicuña, by contrast, is one of South America's great quadrupeds, whose presence on the continent predates the arrival of humans there. After Brack's plans for a vicuña industry became known, there was a public outcry, spearheaded by Benavides and Peru's elite press. 'To offer vicuña chops was and is, in sum, like promoting the eating of panda crackling,' argued Peru's prominent weekly *Caretas* in 1986.

In the course of his career, Benavides had gone after anyone who threatened Peru's wildlife: he attacked Aristotle Onassis when his whaling ship entered Peruvian waters; he threatened to club seal clubbers; to shoot Abdul Reza Pahlavi, a half brother of the shah of Iran, if he killed a spectacled bear; and to confiscate a vicuña fibre poncho given to Fidel Castro. But nothing really compared to his attacks on Brack. Benavides accused him of falsifying population figures to justify the venture's commercial objectives. 'Not even rabbits reproduce at the rate the government claims, much less the vicuña,' he said. He invited in a biologist from Cambridge, Dr Keith Eltringham, to carry out an aerial census, which concluded that there were but a third of the number of vicuñas that Brack had recorded. But it was Benavides's comments on the cull, his ability to conjure Dantesque images of a bloodbath, that were most effective in swaying public opinion. 'They wound the animals through the lung, as their hearts have a good price as a local delicacy,' he said. 'Many wounded vicuña are running with their guts hanging out, and die in agony.'

'The saga of the vicuña', as the dispute later became known, dragged on through the 1980s, becoming more and more personal and embittered. It was ostensibly about the right to kill an animal, but it was from the outset much more than that, a microcosm of divisions within Peruvian society. Wealthy, confrontational and well connected, Benavides was viewed by many Peruvians as an aloof aristocrat, a man who cared more for vicuñas than for the desperately poor campesinos who shared their land with the camelids. Brack, in turn, was represented – quite unfairly – in the press as a mercenary butcher out to monetise a sacred creature. Technically, Benavides won the war: in the 1980s the cull was abruptly halted by the government, and Brack's organisation, the Project for the Rational Use of the Vicuña, was dissolved. But the dispute consumed Benavides, alienating him from conservationists, and he fell out of political favour. In 1991 he was fired from his job as head of Lima's zoo, the Parque de las Leyendas, by President Alberto Fujimori. One month afterwards, Benavides died of cancer in London. His obituary in *El Expreso*, a Peruvian daily, was titled 'El día que las vicuñas lloraron' – 'The Day the Vicuñas Cried'.

There's a detail in the history of the conquest, the story of how fewer than two hundred Spaniards subdued the Incan Empire, that has always intrigued me. In the period before the Spaniards execute Atahualpa, the Incan emperor, Pedro Pizarro notices his attire, a dark brown cloak as soft as silk. Intrigued, Pizarro, then a teenager, asks Atahualpa about the material of which it is made. According to Pizarro, Atahualpa replies that it is made of the skins of vampire bats, harvested in Puerto Viejo and Tumbez, two northern provinces within his empire. 'Those dogs of Tumbez and Puerto Viejo,' Atahualpa reportedly says, 'what else have they to do than to capture these animals so as to make clothes for my father?'

When the Spanish penetrated into Incan territory, they came across huge warehouses piled with goods. They contained items of great value and utility – food, arms, ornaments and tools – but the Spanish were most startled to find vast quantities of wool, cotton, cloth and garments. 'There was so much cloth of wool and cotton', wrote Francisco Xérez on seeing a warehouse in Cajamarca, 'that it seemed to me that many ships could have been filled with them.'

Faced with the abundance of cloth, stored as if it were treasure, the Spanish were reduced to a state of speechlessness. 'I could not say about the warehouses I saw, of cloth and all kinds of garments which were made and used in this kingdom, as there was no time to see it, nor sense to understand so many things,' wrote Pedro Pizarro.

It is not hard to grasp why Pizarro was so bewildered by the abundance of cloth in the Incan Empire. The Spanish came to South America in search of gold, looting temples and fortresses, but they found an empire that was held together by fibre. 'There is nothing strange in the use of prestige objects', writes historian John Murra, 'the novelty consists in discovering that, in the Andean area, the artefact of greatest prestige and thus the most useful in power relations was cloth.' Skilled weavers, the Incas relied on cloth for countless functions, from sacrificial offerings to burials. According to Murra, 'No political, military, social, or religious event was complete without textiles being volunteered or bestowed, burned, exchanged, or sacrificed.'

The Jesuit Bernabé Cobo related that the Incas had two types of fabric: *awasqa*, a rough, thick cloth for domestic purposes, and *kumpi*, a finer fabric that was described as being as soft as silk. But the most coveted materials were those that were wild harvested, often reserved for elites. According to Garcilaso de la Vega, the Incas harvested vicuña fibre in a *chaku*, a practice in which the animals were live-shorn. He wrote that the Incas' hunting party would assemble twenty to thirty thousand people to form an enormous circle around thousands of wild animals. After closing in on them, male vicuñas and guanacos (the large cousin of the vicuña) were killed, but females were shorn and freed. The round ups were carried out on a rotational basis, allowing stocks to recover and the animals' fibre to regrow. When Benavides read these accounts of the chroniclers, he was fascinated. 'Those great conservationists, the Incas, were very wise in the use of their wild animals,' he wrote.

Inspired by these writings, he began to conceive live-shearing as a means by which the animals could both thrive and provide income for local communities without culling. Surrounded by more than two hundred parrots, ocelot kittens and a box turtle, he would lecture journalists on its potential. 'The people of the Andes can shear the wool,' he told the *Los Angeles Times* in 1975. 'We are working out ways to catch the vicuña humanely with invisible nets, shear them and release them again.' He would even take visitors to Lima's zoo, where he would shear tame vicuñas in a field behind his office. 'I want everybody to have a vicuña coat,' he told the *Los Angeles Times* in 1978. 'I just don't want the last male and female vicuña killed.'

In its idealism, Benavides's plan reminded me of Gavin Maxwell's attempt to reimagine an Icelandic tradition, though the Peruvian wanted to transport a tradition through time, some five centuries, rather than across an ocean. Throughout the 1980s Benavides lobbied hard to open up the trade, successfully applying to shift selected Peruvian vicuña populations into Appendix II of the Convention on International Trade of Endangered Species of Wild Fauna and Flora (CITES) in 1987. In 1991, the year of Benavides's death, one of the last hurdles to the trade was removed when President Fujimori signed a decree granting the right of usufruct to the nation's campesino communities.

f the idealist of the *chaku* was Felipe Benavides, then its chief practitioner is Alfonso Martínez, a lawyer now based in Lima. The son of a livestock trader, Martínez could not be more different from the Peruvian aristocrat, the son of a diplomat who dreamed of live-shearing. Born in relative poverty, Martínez grew up in Lucanas, then a small village next to Pampas Galeras, whose inhabitants eked out a living herding cattle or pigs. 'Most of my classmates [at primary school] were five or six years older than me,' he recalled. 'Their parents always sent them to school late because they had to help in the fields.' As Lucanas did not have a secondary school, he went to Lima to study, returning during the holidays to help his family, until he got a place to study law in the capital.

Unlike Benavides, Martínez's involvement with vicuñas emerged not out of any great interest in conservation, but because of his connection to his town. In the mid-1980s, shortly before graduating, he became an advisor for his community in Lucanas. 'I accepted because I had the attitude of helping the community. I admit that I was leftist back then, strongly to the left.' During that decade, he and others in Lucanas began to lobby the state to give them custody of the vicuñas. 'We'll look after them, defend [them], and benefit from them,' he told officials. 'The state was very interested in conservation, the protection of the vicuñas, but with the incursion of Sendero Luminoso, the vicuña was at the mercy of poachers, the army, and Sendero.'

Sendero Luminoso (Shining Path) was a Maoist guerrilla group that seized control of large parts of rural Peru, including the Pampas Galeras National Reserve, in the 1980s. Founded by a university philosophy professor in the late 1960s, the group indiscriminately killed campesinos, political opponents and vicuñas in what the writer Mario Vargas Llosa described as an attempt to 'throttle . . . Peru's cities to death, Lima above all, by allowing no food to reach them.' After so many years of bloodshed, Fujimori's piece of legislation, the gift of the right to shear, was gratefully received by many campesinos, though the actual idea of shearing a vicuña, a tetchy creature that can reach speeds of up to thirty miles an hour, remained perplexing. 'They had no idea how you could capture a vicuña,' Martínez told me. 'When the communities came to Lucanas from the sierra they said: "You're mad. How on earth can we catch a vicuña if it can run one hundred times faster than us?"'

In Pampas Galeras you can find old stone traps, which many believe were used by pre-Incan peoples when rounding up, shearing or killing vicuñas. But in general there is little historical information concerning the Incan *chaku*, no archaeological remains and very scarce written evidence. 'The only material that we find is in the chronicles,' Martínez acknowledged. 'But if you read them it's quite generic. They don't go into any details. So we had to come up with a series of aspects tied to Andean customs.' He suggested rounding up vicuñas with a large rope with flags on it, using hundreds of volunteers; the vicuñas would then be shorn with electric shears, just like the Incas had done, electricity excluded.

The historian Eric Hobsbawm defines an invented tradition as 'a set of practices, normally governed by overtly or tacitly accepted rules and of a ritual or symbolic nature, which seek to inculcate certain values and norms of behaviour by repetition, which automatically implies continuity with the past.' Martínez's *chaku* clearly fit this definition, blending the outlines of an Incan tradition, Andean customs, modern-day technology and his own imagination. As a model, the *chaku* proved highly successful in Peru; hundreds of campesino communities organised themselves into associations to harvest fibre and sell it in the international market. 'On the back of the vicuña is one hundred dollars,' Martínez said. 'If you see a helicopter and the president of the Republic is there and he drops a wad of cash that is going to float away in the wind, what do you do? You run towards those bills.'

The first *chaku* was held in 1993, to great fanfare. It was a big event for the community, the reserve full of government officials, reporters and politicians, including Fujimori. 'There was a lot of idealism because we wanted to link all of this project with Andean identity,' Martínez recalled. Yet any enthusiasm was tempered by the practical difficulties of catching these nimble creatures. The first year they got a mere seven; in later years the *chaku* was even less successful. 'There was one year when we didn't capture a single one. And that was because someone in a key position fell asleep.' I appreciated these difficulties myself when I later visited the reserve and tried to creep up on the vicuñas. As soon as I got within twenty metres of a band of vicuñas, the male would prick his ears, raise his head, let out a high-pitched screech and disappear with the females.

In the northern Arctic regions, the Sami faced similar problems when trying to round up their herds of reindeer. So testing were their round ups that the Sami sometimes abandoned their nomadic ways, building miles and miles of dispersed fences to contain their animals. This allowed them to herd larger numbers of animals, kill them and benefit from a bonanza of meat and skins. Martínez explained how he made the *chaku* more efficient by using a similar technique. In 1996 he and CONACS, the state body that regulated the shearing, started encouraging local communities to build wooden fences around their land to channel the creatures into corrals. Instead of catching a few vicuñas, communities could haul them up by the hundreds. 'When I started, the risks of not capturing were very high,' Martínez recalled. 'Now even a blind man can round up vicuñas in Pampas Galeras.'

The logic behind Chiara Vigo's museum in Sant'Antioco was that the more sacred an object, the more likely its owner was to survive. Commercialise byssus and you'd open up a Pandora's box: *Non si vende, non si compra.* Yet when I read about the case of the vicuña, it suggested a different set of rules: the more fibre was harvested, the more vicuñas there were, protected by the communities that had a stake in their survival. Between 1994 and 2012, Peru's vicuñas more than tripled, numbering almost 210,000 in 2012, the date of the most recent census.

As the vicuña numbers boomed, I noted that their fibre took on a fabular quality, appearing in pro-market studies and press releases as proof of the merits of commodification. If only, the fibre seemed to proffer, we could find a market for all materials from endangered species, then their owners might prosper while providing much-needed income for impoverished communities. Nowhere is this more evident than in arguments about the legality of rhino horn, an illicit commodity that is said to have much in common with vicuña fibre. Like vicuña fibre, rhino horn regrows and can be cropped without killing its owner; it is also subject to huge demand, mostly emanating from China, which has led to widespread poaching. In recent years, a South African businessman named John Hume, who owns some 1,500 captive rhinos and a stockpile of horn, has vociferously lobbied for the legalisation of the trade in live-cropped rhino horn, citing the example of the vicuña. 'I trim [rhino] horn in such a way that it can regrow again,' he told *The Mercury*, a South African publication. 'This is the way to save them from extinction.'

Over the past few decades the vicuña was not the only wild creature to find itself introduced into the marketplace under CITES's approval. Crocodiles are now widely farmed, from Borneo to Australia, their skins harvested for bags and clothing; butterflies are raised in captive conditions, and sold to private collectors or zoos. But the vicuña was different from these cases, offering the possibility that a wild animal could pay its way without suffering unduly or dying. Once a symbol of decadence, the fibre became a talisman offering its own mantra: A *vicuña shorn is a vicuña saved*. Much like Chiara Vigo's museum, Pampas Galeras became a sort of shrine, a pilgrimage site visited by journalists, conservationists and tourists eager to witness the workings of the *chaku*'s logic and spread its message. One year I joined them, too.

During the 1960s and 1970s the reserve of Pampas Galeras was at the forefront of efforts to save the vicuña from extinction. Funded by the German government, it boasted high-quality facilities, including a school for the children of park staff, a research centre, some sixty horses, and vehicles. 'We had a load of people back then,' recalled Reino Joyo, a retired park guard, who started working in the reserve in 1977. 'We went out on armed patrol every day.' It was a bitter struggle to stop the poachers, he recalled. Armed with rifles, they would fly in by helicopter, massacre the vicuñas and return to Bolivia, the gateway to the illegal trade in skins. During firefights it was not uncommon for guards to be wounded or killed; Reino, his voice breaking with emotion, told me that twelve guards died defending the vicuña in those years.

I'd only recently arrived in Pampas Galeras, having taken a *combi*, or minibus, from Nazca, the coastal town. Travelling on the main road between Cuzco and Lima, you ascend several thousand metres in a few hours, the landscape transforming from desert to puna. Unacclimatised, I fell asleep for much of the journey, waking at four thousand metres with a pounding headache to find a flat expanse of grassland covered in the cinnamon hues of vicuñas. Walking through the ruined buildings of the reserve, I found it hard to imagine that it had once been a state-of-the-art centre of conservation. Close to its entrance was a rusted, burned-out truck, its smashed windows forming a mosaic of glass on the grassland.

Crunching the detritus, I headed to the reserve's bridge, a mangled rib cage of shrunken wooden planks, and walked through its ruined lodgings. Roofless and peppered with bullet holes, they had been colonised by viscachas, rabbit-like rodents and vicuñas, their presence indicated by neat piles of dung. A far cry from the 1970s, the park staff today live in creaking old dormitories without electricity or heating; visits from outsiders, apart from long-distance cyclists or passing truck drivers, are a rare event. 'The reserve was one hundred per cent better back then,' Reino said. 'Now it's dead. Dead, dead. If the Germans came now, what would they say?'

Along with bullet holes, you can find clues to the violent past of this place: a hammer and sickle daubed in red paint on buildings as well as letters that spell a surname, Guzmán. In October 1989, Sendero Luminoso fighters unexpectedly appeared in the reserve, looking for its head, Héctor Galván. It was the second occasion Sendero had entered (during the first, in March 1983, they killed two officers of the forestry police), but this time they wished to decapitate its leadership. 'They always tended to kill the chiefs,' Hernán Sosaya, another guard, told me. 'They would kill the mayor, the leader of the community, the president, the governor; in this case it was the same.' Unable to identify Galván, who pretended to be a campesino, the fighters ordered the reserve staff to paint its buildings with Marxist symbols. 'All of us painted,' Reino said. 'We had about forty cans of red paint in the warehouse.'

It is hard to think of a worse place to be in the world in the late 1980s than Pampas Galeras. During that bloody decade, the reserve found itself in the heart of the conflict between Sendero and the Peruvian government, one that would eventually claim almost seventy thousand lives. Once the site of one of the great conservation successes of the twentieth century, the puna became a graveyard, littered with explosives and human remains. 'When Sendero was here, you would find skulls in ditches,' Hernán said. 'They would use heads as footballs.' Why, I asked, did the guards not leave earlier? 'Everyone who works with the vicuña, we call them *vicuñeros*,' explained Hernán. 'When you first set foot in this job, you never stop working until you die. We call it the spell of the vicuña.' In Spanish, a simple suffix – *ero* – may represent a relationship of mastery and control over an animal: a *cabrero* herds goats, *cabras*; a *caballero* rides horses, *caballos*. Yet the *vicuñero* did not dominate, but protected the vicuñas, taking poachers' bullets and holding firm as the puna was drenched with blood. 'We were lovers – *amantes* – of the vicuña,' Reino told me.

I once read about an Andean myth that tells that the Incan emperor wanted to meet a girl from a village who was desired by all the men of the region. Her parents did not want her to meet the Inca and so they transformed her into a vicuña. Bewitched by her beauty, the Inca would visit her every day, until she was killed by his jealous wife, who turned her skin into a dress. I'd think of this story when I spoke with the *vicuñeros*, who used a language to describe these creatures that was more befitting a human being than an animal. Perhaps, I thought, it was a love based on the knowledge of sacrifice, the lives lost in its defence; or the vicuña's inescapable beauty, that it is a wild creature, and yet its eyes and delicacy elicit a response that one might expect of a domesticated animal, selected to mirror human desire.

Within a week of the attack in 1989, Sendero carried out a full-on assault on the reserve at night, burning buildings and spraying them with bullets. This time no one stayed. The staff fled into the grassland, including Galván's wife, Marta, then eight months pregnant with her third child. 'I headed for the hills with my two small girls, my tummy and another girl,' she told me. 'As it was so dark, I kept on falling over.' The following day, the Sinchis, the government's death squads, arrived to rout Sendero, torturing or killing anyone whom they suspected of having a connection with the guerrillas. 'In '87 they killed my mother and brother, who was only three, with dynamite,' Hernán said. 'We don't know who did it, but the state gave us an indemnity of 10,000 soles [about $3,000], nothing more.' After the arrival of the Sinchis, regular government soldiers took over the reserve, cannibalising its facilities. 'The vicuñas were targeted by poachers because there was nobody here,' Hernán said. 'The puna was emptied.'

When I visited Lima or read accounts of the *chaku*, I thought that it was an idealistic experiment, a testing ground for our relationships with these wild animals. But the more time I spent in Pampas Galeras, the more I appreciated that it was a gauze, a means to heal the wounds of those bloody years that had led to a total collapse in the conservation efforts of previous decades. Less utopian dream, the *chaku* was a suture, a last-ditch resort to find value in these wild creatures that were nothing but a nuisance to local people. 'Who helped the poachers?' asked Manuel Cabrera, another retired *vicuñero*. 'The *comuneros*! . . . The hunters came along, they would give you ten or twenty soles [three to six dollars, for help]. They would take everything, apart from the meat, which would stay.'

Almost three decades on from the capture of Guzmán, the *chaku* is now a well-established part of life in the puna. The week that I visited would be the twenty-fourth since custody of the animals was handed over to campesino communities; the reserve, normally silent apart from the vicuñas' cries, was a hub of activity: the park staff were busy, cleaning bathrooms, stocking up on food and opening dormitories to host hundreds of guests, who would include politicians, a South Korean film crew, the news media, Inca enthusiasts and an Englishman who had overstayed his welcome.

On the morning of 24 June we spread out across the puna in a line, holding a long rope with coloured plastic flags that crackled in the light breeze. It was a day so bright that the grassland, which had initially seemed barren and desert-like, suddenly revealed its varied textures: the white wisps of the beards of small cacti, known as *viejos* (old men); the neat circular deposits of vicuña droppings. At the command of the park guards we started walking forwards.

As a boy, I once went stalking in Scotland as part of an annual deer cull. In a pair of borrowed itchy tweed trousers, I shadowed a couple of gamekeepers skulking a large herd of deer in the highlands. Some twenty years on and I remember little of that day, except the sheer discomfort of crawling through heather, the dense fogs of midges and the command to be still, to be absorbed by the landscape. Here, sweeping through the puna in bright Gore-Tex, we were raucous, more festival-goers than hunters. Like drunkards, we staggered through the grassland, flushing out its residents: vicuñas, viscachas, foxes and flightless juvenile Andean geese. Above, flocks of birds moved away from us as if a great barrier stretched vertically from the rope to the wisps of cloud overhead.

The vicuña is built for altitude, its body perfectly crafted to transport scarce oxygen molecules at four thousand metres. At between 0.7 and 0.9 per cent of its total body weight, its heart is roughly double that of similarly sized mammals. This powerful organ pumps oxygenated elliptical blood cells through a dense network of capillaries to its limbs. We enjoyed no such advantages. In the altitude we grew irritable, the cold air burning our throats. We croaked and wheezed and swore. We tripped up on the bearded cacti, bits of rubbish and camelid skulls and spines, ignoring the sight ahead of us. In the distance a nutmeg patch emerged, a group of about thirty vicuñas surging forward to test the line. 'Shake, shake!' cried the guards.

Approaching individual vicuñas, I was always conscious of their fragility, the delicacy of their limbs and slender necks, but now, faced with a group bolting toward us, it was I who felt fragile. We shook the rope, flapping the plastic flags with the vigour of a political rally. The collective noun for vicuñas is *tropilla*, or band, but many in Pampas Galeras talked of *olas* of vicuñas – waves. Watching them now, I could understand how apt this word was. As they ran past us toward the corral, their lithe forms seemed to roll, their heads locked in place while their backs undulated, as if an impulse were being transmitted among them. To my relief they banked to our right, heading down the line, before turning again towards a fence that led to the corral in the distance.

In Sardinia, across Sant'Antioco's lagoon, lies another small island, San Pietro, once famous for its annual tuna harvest, known as the *mattanza*, or killing. Until recently the island's fishermen would head out in their boats to intercept hundreds of migrating tuna. Relying on a system of nets, often more than a kilometre long, the fishermen would channel the tuna into different *camere*, or rooms, before raising the nets and harvesting their catch. That image comes to me when I remember the vicuñas thrashing in the corral, their long necks buckling against the nets. Like those fishermen at San Pietro, we had not stalked or hunted these creatures, but channelled them and anything else in their vicinity. At the fringes of the pen I saw a fox and two Andean geese, an unfortunate bycatch, trying to dodge the branches of vicuña legs. Within minutes, a small dust cloud frothed up, concealing the vicuñas' movements, but their sweet, smoky smell lingered in the air.

The Spanish chronicler Pedro Cieza de León, who lacked Garcilaso's romantic images of Incan conservation, wrote that, after rounding up wild animals, 'certain Indians entered the enclosure with *ayllos* [lassos] which are used to secure the legs, and others with sticks and clubs and began to seize and kill. . . . They [the guanacos] tried to escape by spitting into the faces of the men and rushing about with great leaps. They say that it was a marvellous thing to hear the noise made by the Indians in catching them, and to see the efforts made by the animals to escape in all directions.' Unlike the Incas described by Cieza, the vicuña shearers of Pampas Galeras do not shed any blood; their job is far more demanding, more nuanced, than simple slaughter. Shearing is an ancient technique dating back at least to the Iron Age, but it normally involves the restraining of domestic animals rather than wild creatures that fear our presence; to shear a vicuña, to remove a part of its body, seemed as sensible as riding zebras or herding bison.

Intrigued by the strangeness of this image, I joined the shearers in the early hours of the following morning. I clambered into a

Japanese truck laden with ropes, metal stakes, electric shears and the bodies of twenty shearers, slotted in the back like pieces of a backgammon board. The head shearer, Canchito, a taciturn former livestock herder with a blue baseball cap and a large gold watch, turned to me. 'What do you capture in your country?' he asked. 'Tourists come from all over. They told me that in their country they capture elephants, zebras, other camels. What do you capture?' Fearing my answer would disappoint, I remained silent.

We drove onto the dirt track of the puna, bouncing up and down on the wooden boards of the truck and regretting the vast quantities of *calentito*, a local drink made with hot water, lemon juice, rum and cane sugar, that we had drunk the previous night. When we approached the corral, the vicuñas seemed calmer than they had been the previous day, gently swirling around the rock in the corral's centre, a cinnamon liquid. I was relieved that there was no sign of the previous day's bycatch, the fox and geese, but overhead I noticed another visitor, a vast Andean condor, perhaps hoping that there might have been a casualty overnight.

One by one the men moved into the corral, grabbing the vicuñas by their tails and necks. When held, the vicuñas kicked, wailed and bit, their slender legs bucking frantically into the air. Every so often, a vicuña would hit a man's genitals, leaving him rolling in the dust, groaning and swearing. As each vicuña was hauled out, Canchito placed his hand on its back to measure its fibre, the bright gold of his watch clashing against the darker, more muted sponge. If a vicuña had been shorn recently or was a juvenile, it was immediately re-leased. Other vicuñas were taken off to one side to receive a special treatment for mange, a skin disease caused by mites (it occurs in humans as scabies). A sorry sight, these creatures' bellies and haunches were crusty and dry, almost reptilian. The men pinned them on their backs as Norma Bujaico, the community's vet, carefully greased them with motor oil to kill the parasite, covering their bellies and haunches with black treacle.

I watched Canchito's hand disappear into the fibre of a headlocked vicuña. Ripe for shearing, it was hauled forwards by two men, its outstretched legs tied and bound to metal stakes with a rope. Once trussed up, its neck locked in place by its own legs, the beast lay motionless, inert as a carcass; only its blinking eyes, flaring nostrils and the occasional moan suggested life. On cue the sound of a mechanical rasping was heard; a man, the designated shearer, appeared, wielding an electric razor to carry out the final act.

It felt like we were preparing for a sacrifice, the execution of a wild animal as part of some ritual; I half expected the man to put the razor to the vicuña's neck, slitting its throat, but he swiftly passed it over the animal's slim body, scooping off the ultrafine fibre as if it were foam. Within twenty seconds, the vicuña was nude, its fleece piled up in large spongy clumps, leaving behind a series of tram lines. The fibre was shoved into a plastic bag, the ropes were untied and the animal, shell-shocked, sprang up and made its escape.

I lifted one of the bags and removed a clump of fibre, relishing its sweet smell. Fresh off the vicuña, it was still warm, and I was tempted to unravel it and wrap it around my neck. Touching it, I could understand why Europeans and Americans so coveted vicuña fibre, and spoke of it in almost magical terms. 'It is thick and bushy, extremely fine, soft and silky to the touch; and possesses an extraordinary gloss and lustre,' wrote the Bolivian intellectual Vicente Pazos Kanki. My hands warmed by the fibre, I was overcome by a rapacious desire and half contemplated the possibility of slipping it in my rucksack while the shearers were occupied, but I also felt mildly discomforted. Unlike eiders or canids, vicuñas do not moult, retaining their fibre year-round to protect them from the cold; was the warmth that I felt the counterpart to the animal's coldness?

The vicuña cannot speak, but biologists have tried to understand what these creatures feel when they are rounded up and shorn, measuring their rectal temperature, heart and respiratory rates, creatine kinase levels and plasmatic cortisol concentration. According to Cristian Bonacic, a Chilean ecologist who developed guidelines for live-shearing of camelids back in the 1990s, the round ups and shearing can cause acute stress in vicuñas, but so far there is little evidence that the *chaku* is responsible for a notable increase in mortality: the question of whether the stress induced by the *chaku* is acceptable remains one of personal ethics rather than wildlife management.

I watched the vicuñas head off into the distance, their backs blending into the grass of the puna. Tonsured for market, they cut haggard forms, their grizzled white bibs hanging down like beards. One vicuña simply stopped dead in its tracks after bolting off, unable to find its family group. Pivoting on its hind legs, it jumped left and right, and scanned the horizon for its family. It must have repeated the motion a dozen times like a stuttering second hand, a clockwork camelid. Norma assured me that they would eventually find their family groups, returning to them as if little had happened, just like Iceland's eiders.

Was the *chaku* a useful model? Could rhinos be cropped of their horn, saving them from extinction? After several weeks in the puna, I was not so sure. In an interview with *National Geographic*, Bonacic points out that – unlike rhinos – the population of vicuñas in many parts of South America had largely recovered by the time the trade in their fibre was legalized; he says that the legal trade in vicuña fibre has led to more poaching, not less, and that there are many differences between the *chaku* and horn-cropping, rendering any comparisons difficult. As a weapon that is important in rhinos' reproductive lives, horn serves an entirely different function from that of fibre. Unlike vicuñas, rhinos must be tranquillised while their horns are removed with an electric saw.

In any case, some biologists have questioned whether the *chaku* has really been good for the vicuñas, even though their numbers have swelled. Jane Wheeler, an American archaeozoologist who was the first to unravel the evolutionary history of camelids, told me of her concerns about the sanitary and genetic consequences of the *chaku*. Each intervention, she said, from the fencing of wild animals to their indiscriminate injection with antibiotics, compromises the genetic variability of the vicuñas, making them more vulnerable to disease and climate change. A particular concern is the use of fences, now widespread throughout Peru, and their effects on the animals' social structures. '[In general, in a vicuña band there is] one male for every five to seven females,' Jane explained. 'The male expels the offspring when they are ten to eleven months old. The males that are expelled go off and join bachelor bands. And the females join other family groups which are not genetically related. Ideally the whole structure prevents inbreeding and strengthens genetic variability which strengthens the chances of survival.'

In *The Variation of Animals and Plants Under Domestication*, Charles Darwin wrote of his admiration for the Incan *chaku*, calling it the 'most curious case of selection by semi-civilised people, or indeed any people, which I have found recorded'. Like good gamekeepers, the Incas pruned the weakest animals during their round ups but saved the strongest animals to the benefit of the species. This present-day *chaku* could not be more different. Fenced in, shorn as sheep, rounded up, injected, greased as machinery, traded as cattle, the vicuñas were now treated much like domestic livestock, assets to be protected. In the process these creatures had begun to change: weaker animals now survived, animals from different groups were mixed together and interbred, and fences distorted their normal breeding habits. In some communities vicuña populations are exploding, placing great pressure on habitats already strained by desertification and climate change. 'It's a nightmare,' Jane told me, shaking her head.

n the early nineteenth century, Empress Josephine, the wife of Napoleon, requested that Charles IV, the king of Spain, send over a batch of camelids from South America for her residence at Malmaison. In accordance with her instructions, some thirty-six creatures left Buenos Aires on a frigate destined for Spain. From the outset their journey was hampered by difficulties. During the voyage, the frigate was attacked, possibly by the British, and part of the supplies were lost. '[The animals] were afterwards fed with potatoes, ears of maize, hay, and bran', reported Don Francisco de Theran, a Spaniard who later cared for the animals. 'While they had potatoes they did very well; but as soon as this food was no longer to be had, they became diseased, and many of them died.'

During the centuries after the conquest, the Spanish brought pigs, sheep and horses to the New World, displacing flocks of camelids, which they believed to be inferior. It charmed me that the empress had reversed the flow, introducing these curious creatures to Europeans, who referred to them as '*ovejas peruanas*', or 'Peruvian sheep'. By the time the frigate arrived in Cádiz, only eleven of the thirty-six camelids had survived, of which at least two were vicuñas. Worse was to follow; once the starved camelids landed in Spain, they found themselves at the centre of a full-blown political crisis. According to the Scottish zoologist James Rennie, they docked just at the start of the May 1808 uprising, during which the Spanish populace turned against Prime Minister Manuel Godoy, the man who had brokered the arrival of the camelids. 'In hatred of their late minister,' writes Rennie, 'the populace were about to throw the llamas into the sea'.

Yet no camelid, we are told, was tipped into the Atlantic; instead they were rescued by the governor of Cádiz and entrusted to Theran, the intendant of an acclimatisation garden in the town of Sanlúcar de Barrameda. An enthusiast of exotic species, Theran took to the animals, documenting their habits and even making himself a hat from camelid fibre. They apparently lived in the garden under his care and were later placed under the protection of Napoleon's general Marshal Soult during the French occupation of Andalusia. Given their rough crossing, the political instability in Cádiz, and the disruption of the French occupation, perhaps it is no surprise that Empress Josephine's exercise did not end well. According to one later observer, the camelids perished within three years, leaving no offspring. But Theran believed that the episode had set an important precedent. 'After the experiment . . . ', he wrote, 'I am convinced that the acquisition of the vicuña, as a domestic animal, will be one of the most interesting conquests which industry has made of the animal creation.'

Over the next century, the 'acquisition of the vicuña', as Theran put it, would become an obsession among certain Europeans. Enchanted by its silky wool, they wrote of it as if it were a prize to be captured, a summit to be conquered, or a pole to be reached. If only vicuñas could be brought under control, bred in captive conditions, then the treasure of their wool could be secured. The most ambitious advocate of domestication was a British political journalist named William Walton. In the early 1800s, Walton proposed a scheme whereby the 'Indians' of South America could domesticate vicuñas in return for an exemption from the *mita*, the Spanish land tax. He envisaged that vicuñas could be brought over to Europe to start a new industry. 'The mountains of Wales,' he wrote, 'and those of Cumberland, where there is abundance of moss, would constitute a congenial climate, as well as almost all the downs on our Northern sea coasts.'

In the accounts of men like Theran and Walton there is a great confidence that the vicuña could be brought under control. 'It is a fact,' wrote Walton, 'of which we have the evidence within ourselves, that the Vicuña, when taken young, can be tamed and domesticated, in such a way, as to become an inmate of the cottage, and serve as a companion for children to play with.' But the scenes Walton envisaged never materialised; no 'Indians' in Peru tamed and domesticated the vicuña; no herds of vicuñas ended up rolling through the Alps, the Pyrenees or English grasslands. Although vicuñas can be tamed, particularly when young, no one in recent centuries has managed to get them to breed in captivity, except by crossing them with alpacas. By the 1950s many researchers had concluded that vicuñas, like zebras, were destined to remain in a state of 'perpetual wildness'.

Watching Canchito and his men try to catch vicuñas, I could understand why it was that so many believed the vicuña was destined to tread a different path from our own, forming two threads that would never cross. And yet such beliefs about the impossibility of the domestication of these animals later proved to be deeply flawed. In the 1980s, Jane Wheeler examined about a tonne of camelid and deer bones from Telarmachay, a rock shelter in the central Andes that had been inhabited by humans for eight thousand years. Comparing these bones with present-day camelids, she concluded that the human inhabitants of the rock shelter had in fact domesticated the vicuña six to seven thousand years ago, eventually yielding today's alpaca, one of South America's two domesticated camelids; her findings were later confirmed through DNA.

Wheeler's research overturned the widespread belief that pre-Columbian peoples had been incapable of or uninterested in domesticating the vicuña: she proved that men and women living in a rock shelter thousands of years before the arrival of the Spaniards had achieved something that 'modern' humans had struggled to achieve. 'No one wanted to believe it,' she said. 'People have this tendency to think that the native population, the "Indians," were incapable of doing something like domesticating the vicuña or the guanaco. [. . .] The idea is that if we – modern people with all of our knowledge, science, everything – cannot domesticate the vicuña then those "stupid ignorant Indians" never could have done.'

I found something compelling in Wheeler's research, the way it asked hard questions about progress, technology and our own familiarity with the wild. Unlike those who catch vicuñas today, the men and women of Telarmachay had followed the vicuñas as the glaciers receded, hunting them and living off their flesh. 'To be successful at hunting,' Wheeler told me, 'you have to learn what an animal is like, how it adapts; you have to know everything about it, where you are going to find it, how many males, females, what have you – it requires very intricate knowledge.'

I counted the number of claims that piled up on the bodies of these vicuñas. The *comunidad* and the firms wanted their fibre, a raw material to sell in the marketplace; the politicians needed a nationalist prop, a means to bind these remote communities to the state; the tourists sought an encounter with a wild animal. And I wanted my story, another object. We were like a line of creditors, waiting for our pound of flesh from these wild creatures, calling in their debt, the price for their wildness. No blood was spilled, no flesh was carved, no shot was fired, but it felt as if something had been sacrificed in Pampas Galeras. I feared that, asked to be spectacles, national symbols and carriers of fibre, the vicuñas might simply collapse under the weight of our competing desires, their bodies buckling. And yet these claims meant that they had proliferated in a world dominated by a species that is gradually driving the wild to destruction. Left alone, they might have shared the fate of the guanaco, their wild cousin, which is now critically endangered in Peru.

Such questions did not figure prominently in the lives of the shearers, who lived firmly in the present, their gaze set on the needs of their families. I had heard much about the poverty in Ayacucho, Peru's poorest province, but it was not until I saw the shoes of the shearers that those abstract figures on incomes became clear. Glancing at their footwear, a collection of old trainers and army boots, I found few that were not torn or full of holes. After shearing the vicuñas, they sat down by the corral, exhausted after two days' work at high altitude, and opened a couple of crates of beer, its smell mingling with the smoky reek of the fibre. The men drank solemnly, passing a plastic cup around as if it were a ceremonial beaker, while the *chaku*'s president, a former NCO in the Peruvian army, entertained them with a series of well-worn jokes. 'Eight females for every male,' he roared. 'Fuck, would I like to be a vicuña!'

It's a journey of several thousand miles from the reserve to the fibre's destination, Loro Piana's factory in Roccapietra in northern Italy. First the fibre heads to Lucanas, a small town that lies farther north along the winding road to Cuzco. In Lucanas it is cleaned, its long guard hairs plucked out by local women, and weighed, then taken under armed escort to a warehouse in Lima and handed over to Loro Piana, which has the exclusive right to all the fibre produced by the community. It's often said that consumers know little about the origins of the products they buy, but the reverse is also true: although the shearers knew the sale value of the raw fibre, they had little inkling of its ultimate price. At each stage of production, the fibre loses weight but gains value, so that by the end of the journey $200 worth of fibre becomes $10,000 in stores in New York, London and Milan. When the fibre hits the shops, adorning mannequins and models instead of vicuñas, it possesses no smoky smell, no reek of camelid, nor the guard hairs; a silky softness replaces its sponginess, as one would hope, given that Loro Piana offers a vicuña parka for $26,495.

I followed the fibre on the first leg of its journey, taking a *combi* down to Lucanas. I always felt terrified sitting in these small vans, wedged next to a campesino, chicken or child. At such high altitude, the *combis* groan along the steep winding roads, dwarfed by ten-tonne trucks from Cuzco and Lima, whose drivers, I knew from first-hand experience, subsist on little more than several hours of sleep, coca leaves and the melancholic music of Ayacucho. It was not surprising that locals crossed themselves during the journeys, eyeing the obligatory rosary hanging from the rearview mirror, its angle matching every shift in gradient, as well as the flowers and memorials that lined the road.

I preferred to set my eyes on the far distance, hoping for a glimpse of a guanaco. It is rare to spot these reclusive creatures, despite their large size. Unlike the vicuña, the guanaco numbers no more than several thousand in the entire country. On that trip to Lucanas, I didn't see any guanacos, but the search for them allayed my anxiety. As I descended towards Lucanas, my headache eased, I stopped shivering and I began to spot eucalyptus trees and cattle, signs of decreasing altitude. I got off the *combi* in the middle of Lucanas, walked past the *comedores*, adobe houses, a church with a corrugated iron roof, a museum full of pre-Incan artifacts, a school, and stopped in front of a concrete monument, its paint peeling. 'Welcome to the capital of the vicuña', it read.

The inhabitants of Lucanas identify with the Lucanas or Rucanas people, whose ruined dwellings occupy the hill that overlooks the town. Conquered by the Incas probably sometime around the late fifteenth century, the Lucanas were reported by Spanish chroniclers to have acted as the litter bearers of the Incan emperor, noted for their 'even pace' and their light blue livery. When the Spanish came, the people of Lucanas, like all the groups within the Incan Empire, were divided up among the Spanish into *encomiendas*, royal grants, and forced to pay tribute. As with the *chaku*, much of the history of this town is hard to reconstruct, though fragments about the lives of the Lucanas can be gleaned from Pulapuco, the ancient ruins above the town, which hold huge symbolic meaning for its current residents.

More recent history of Lucanas is also hard to reconstruct, but for different reasons. On All Souls' Day 1987, Sendero came into Lucanas, burning shops and killing. 'We saw corpses in the morning in Lucanas', said José Héctor Quispe Mitma, the town's mayor, who was then a twelve-year-old boy. Soon afterwards, he said, the Peruvian government sent in the army and the Sinchi death squads to rout Guzmán's forces. 'Many people left for the west', Héctor said. 'Lucanas was practically a cemetery.' It took years for the peasant communities, the *rondas campesinas*, to drive out Sendero, he said.

You cannot see any bullet holes or the outlines of Marxist insignia in Lucanas, as you can in Pampas Galeras, but the wounds in this town still feel fresh, the graves freshly dug. The first time I visited, its inhabitants were glued to their aged television sets, their screens crackling with images of terrified Parisians. It was the week of the Bataclan attack in France, and news agencies relayed images of the shootings in the nightclub. Few in Lucanas knew where France was, or could understand the language of the victims or grasp the motivation behind the attacks, but they understood the essential meaning of terror. Despite a widely publicised inquiry into the Peruvian conflict in 2003, many in Lucanas are still waiting for compensation and a boilerplate certificate from the government acknowledging the murder or maiming of a family member.

While the central government has not yet settled all the claims, it made another offer to the community of Lucanas: the right to benefit from the vicuña. Before 1991, no one in Lucanas had looked at the vicuña as a natural resource; if anything, the camelids were a nuisance, competing with cattle in the reserve. But now, at the stroke of Fujimori's pen, base metal had turned to gold, and Pampas Galeras, home to a huge population of vicuñas, had all the promise of a freshly discovered mine. In gold rush towns, one may find hints of the wealth that has been extracted from the earth: large town houses, packed watering holes or newly acquired clothes. But as I walked through Lucanas's square, there was little sign of any riches, the wads of cash Fujimori had flung from a helicopter in 1991. Its cinder-block houses, the dust of its unpaved square, its small eateries serving chicken-foot soup to truck drivers, its small shops offering a uniform selection of domestic goods, all spoke of poverty rather than the excesses of a boomtown. Many inhabitants still live without electricity or running water, sleeping under rough ponchos of sheep wool rather than vicuña fibre.

'Fifteen to twenty years of production', Martínez told me when I saw him at his house, an elegant lime-green building close to the main road. 'What does Lucanas have? Physically speaking? You have already seen the office, the school hall, tractor, and two pieces of heavy machinery. . . . That's it. There is nothing else.' Since the start of the *chaku* in the 1990s, Lucanas has remained one of the largest producers of fibre in Peru, shearing between 200 and 700 kilos per year, with prices varying between around $200 and $500 per kilo. 'They never distributed the money to the *comuneros*. Only once, maybe six or seven years ago. A hundred dollars was given to each *comunero*, that's it. The only money the communities ever got was from the cull,' he said, referring to Brack's initiative in the 1980s.

In my old job in financial investigations, I used to try to figure out where money went in large transactions. I'd sift through emails, bank statements and interview transcripts, looking for a clue that would open things up, an account number or a name on a document. When it was solved, I'd get a shiver of euphoria. But I found no such closure in Lucanas. All the people in the town had their own views as to where the money had gone, why it was that the vicuña had not brought prosperity: the international fashion companies that didn't pay a fair price for the fibre, the poachers who killed the vicuñas, the high capital costs of fences, middlemen in the supply chain, interest payments on loans taken out by the community, corruption and outright theft. No one, it seemed, knew for sure, the only indisputable fact being that the promised wealth from the golden fibre had proved an *'espejismo'*, a mirage.

Since the start of the *chaku*, the sale of fibre was managed by a body called the Sociedad Nacional de la Vicuña (SNV), the National Vicuña Company. In exchange for a fee from fibre producers, the SNV negotiated the sale of fibre to international firms, such as Loro Piana. In theory the arrangement made sense for the communities, increasing their bargaining power in the international market, much as marketing boards do for other commodities, from wool to oil. But in practice the SNV was riddled with problems: communities complained that they never received the payments they were due. In 2004 it fell apart, and since then individual communities have negotiated directly with international firms or through middlemen.

Martínez placed the blame at the door of the communities. 'There is no adequate management', he said. 'If you go to the assembly in Lucanas, you will see how the machetes cross.' I did as he suggested, sitting for a couple of hours in a meeting of the *comuneros* to decide the terms of a contract with Loro Piana. It started amiably enough, but within half an hour it had disintegrated into a series of shouted accusations as the vicuña became a rod for every grievance, infidelity or envy. 'I used to consider myself a communitarian', Martínez told me. 'Now I say, "Either you change or you disappear." . . . The way the communities are now, they do not deserve to survive.'

With little to tempt the merchant, apart from metals from its mines, much of the altiplano and puna has remained isolated from markets. Martínez saw the story of the vicuña as part of this history of isolation. Rooted in the past, in the rhythms of subsistence agriculture, the communities had proved unable to handle the demands of this high-value resource and its influx of capital. Sitting in the meeting, he schooled his neighbours on the workings of international firms. 'They have accountants, they have lawyers!' he shouted at them, as if they were school students. He urged them to break the structures of the *comuneros*, incorporate private firms, and start selling their produce, from eucalyptus to local cheese to vicuña fibre. 'I tell them [the community] that they have to develop or they will die.'

In Martínez's view, history was linear, a journey from the primitive or 'idiosyncratic' towards the advanced instruments of capitalism. As one might speak of fences, corrals or electric shears, he spoke of capitalism as a technology, the only means by which Lucanas could develop. 'I have to admit that capitalism has its instruments – and they are the ones that work here . . . My romanticism says that socialism will follow capitalism, but only when humanity changes.' Others in Lucanas told me a different story. Some saw Martínez as an engineer, and less as a prophet, altering the landscape to suit his own commercial interests, concentrating wealth rather than distributing it ever since he lost his job at CONACS, the state regulator, in the 1990s.

'When I left CONACS it was the worst time for me,' he recalled. 'I had a small scallop business . . . in the Paracas Reserve. There was a time when all of my shells died; all of my capital that I had invested went, and this was a serious problem.' He went through a divorce and needed to support his son, who was in school. 'I came and offered my service to Lucanas. I didn't even say that I would charge them that much.' Lucanas, he said, declined, and instead he set up his own firm, Almar, which became the main intermediary in the fibre market, undermining the monopoly of the SNV before its collapse. 'After I got some capital I started buying fibre.'

Today Martínez is among the biggest players in the fibre market. Since establishing his company he's diversified his activities, representing Loro Piana in Peru, trading fibre on his own behalf, and striking up production-sharing agreements with local communities. Versatile and hungry, he appeared to work for everyone in the fibre market – everyone, that is, except the community where he grew up. When I found him in his home, he looked exhausted, his eyes heavy with bags. He was doling out cash to his shearers, young men from Lucanas who travel to different communities to shear their vicuñas. As he walked through Lucanas in his smart shoes, blazer and pressed white shirt, he seemed a stranger in his own town, itself steeped in Marxism. Like the market, he sharply divides opinion in his town, arousing admiration and distrust in equal measure. 'People either love me or hate me', Martínez told me with cultivated disregard. Tired of conflict, he now avoids the community's *chaku*, shunning the event he helped create.

It was hard to pick apart Martínez's history, the allegations and counter allegations that swirled around him, the endless charges and insults. But one thing remained clear: The free market had triumphed in the vicuña trade, challenging old structures based on cooperation that had once defined life in Lucanas. While state or semi-state bodies like the SNV and CONACS had collapsed, Martínez's firm flourished, now controlling much of the trade in Peruvian fibre. In 2000 the Peruvian government signed a piece of legislation, Decree 053, which allowed private landowners to shear

their vicuñas, just like the campesino communities. Far from a campesino tradition, shearing is now open to anyone willing to buy land
in Peru. Indeed, in 2008, Loro Piana purchased its own tract of land
in Peru, and can now shear vicuñas without the assistance of the
communities upon which it had historically relied.

In his essay on Incan cloth, the historian John Murra notes that
whenever the Incas conquered a people, they would bestow gifts of
cloth upon the defeated. The chroniclers viewed this as evidence of
the Incas' generosity, their 'campaign of peaceful penetration', but
Murra saw these gifts differently: 'The compulsory issue of culturally valued commodities in a society without money and relatively
small markets can be viewed as the initial pump-priming step in a
dependent relationship, since the "generosity" of the conqueror ob-
ligates one to reciprocate, to deliver on a regular, periodic basis, the
results of one's workmanship to the Cuzco warehouses.' Perhaps, I
reflected, Fujimori's gift of the vicuña had something in common
with that of the Incas. Viewed as an act of generosity, the fibre had
quietly proved a garb for the advent of neoliberalism, stealthily
undermining communal or cooperative structures. 'It's clear that
the vicuña was privatised', Martínez told me.

One day the vicuña may end up resembling any other agricultural commodity: already it is large private landowners, benefiting
from economies of scale, who produce the most fibre in Peru.
Martínez predicts that the traditional *chaku* will probably disappear,
replaced by specialist shearing teams that can process high numbers
efficiently. The one remaining barrier to consolidation of the market is the prohibition on the trade of juvenile vicuñas, but Martínez
hopes to overturn this, boosting productive efficiency. Not that
many in Lucanas care anymore. 'No one dedicates themselves to
agriculture; no one dedicates themselves to the vicuña', the retired

vicuñero Manuel Cabrera told me. 'Mining is more attractive. In fifteen days in the mine you can earn what the community earns [from the *chaku*].'

I walked up the hill that overlooks the town to Pulapuco, the site of the pre-Incan settlement of the Lucanas. I wandered around its stone ruins, the traces of this ancient settlement, which had been integrated into the Incan Empire. It was getting late; the sun was setting, bathing the town's corrugated iron roofs in glowing red. By now the vicuñas would be settling down in the higher reaches of Pampas Galeras. As this town changed so visibly, its agricultural structures replaced by those of the market, so too did the vicuñas, silently and near imperceptibly. Gradually, perhaps, their fibre was thickening, just as that of their domesticated cousins had after the conquest. One day, perhaps, we will domesticate the vicuña, repeating an act undertaken thousands of years ago – not for want of warmth or flesh, but for our own curious relationship with the wild and its materials.

TAGUA

Before I left Peru, one of the shearers gave me a gift, a small tuft of vicuña fibre about the size of a coin. A fraction of the size of the pastor's eiderdown, it was weightless, but it felt heavy in my rucksack. Under Peruvian law, it is illegal to remove raw fibre from the country without a permit; the legislation is intended to stop the export of fibre from dead animals, but it would apply as much to my live-shorn tuft as to a poacher's bullet-holed skin. Playing with the fibre in my hand, I agonised over whether to slip it in my suitcase, my own good sense trammelled by its hue and texture, its associations with the vicuñas and their *encanto*, their spell.

Although the fibre does not now sit in my cabinet, I did not need an object to remind myself of the vicuña when I arrived back in the UK. Reading the press in April 2017, I learned that South Africa's Constitutional Court had lifted a moratorium on the domestic trade of rhino horn. Within several months, John Hume, the South African rhino breeder, organised the first legal auction of horn from his captive herds. 'No longer will rhino need to be killed for their horn,' he declared. 'No longer shall the supply come exclusively from dead rhino. From this day live rhino shall become more valuable than dead rhino.' Conservation groups vigorously disagreed, arguing that a legal market would only lead to a surge in poaching.

After so much time in the puna, I too was alarmed by this

development, preferring Chiara Vigo's *Non si vende, non si compra*
over *A vicuña shorn is a vicuña saved*. As Felipe Benavides might
have said back in the 1950s, 'Rhino horn looks best on rhinos.' And
yet part of me was intrigued by the availability of this illicit good.
Whenever I visited old English country houses, I'd make a habit of
looking for billiard tables to find elephant-ivory balls, their colours
now faded after years of play. There is an obscenity to the triviality
of a billiard game, one that requires the death of an elephant, but I
also found great beauty in these spherical objects, the contours and
feel of ivory. Like an obsessive I'd furtively touch them, balancing
them in my hand, and then feel slightly guilty afterwards. A certain
shame came over me and I took refuge in another material, which
satisfied my craving for ivory's texture.

n the mid-nineteenth century a notice appeared in a directory from
Portland, Maine, advertising the material as 'the greatest curiosity
of the age: Nature and Art Combined'. Creamy white, dense and
smooth, this curiosity – commonly known as tagua – could have
passed for bone, antler or ivory. Its cellulose, arranged in concen-
tric circles, was similar to the ivory of hippopotamus teeth. Back in
the nineteenth century, only a dab of sulphuric acid could have
distinguished tagua from ivory, for on contact with the acid, tagua
cellulose will turn pink, whereas ivory remains white. Even today, it
is hard to distinguish the two. Fourier Transform Infrared Spectroscopy, a form of chemical and infrared analysis developed in the
1950s, is the most effective test.

The advertisement described the curiosity as 'ivory that grows
on trees' rather than on an elephant or hippopotamus. Tagua is
in fact a nut from a palm – *Phytelephas* (literally 'elephant plant' in

Ancient Greek) – found mostly in north-western South America. In this case, it's likely that the curiosity came from one particular species of this palm, *Phytelephas aequatorialis*, which is endemic to western Ecuador. Growing up to ten metres high, it has large feathered leaves, similar to those of a coconut palm, which spring from its thick trunk. The fragrant blossoms of the female palms develop into fruits, which, clustered together, eventually form club-like burrs or infructescences, each about the size of a human head. Within their fibrous, warty walls are twenty to forty tagua seeds that over a year turn from a milky liquid into balls the size of hens' eggs, which are almost as dense as copper. Once ripe, the burrs fall to the ground, bursting open. Wild hogs, squirrels and other rodents rush to devour the thin, oily orangey mesocarp that covers the nuts.

In the mid-1800s, when European rubber gatherers first came across the burrs lying on the ground in northern Ecuador, they called them 'negritos', curiously believing that they resembled 'in form and color the miniature head of a Negro.' Initially, it must have been hard to know what to make of tagua; the rock-solid mature nuts cannot be eaten or easily burned as fuel. According to one tagua producer, German merchants first used them as ship ballast, a suitable replacement for sand, on which they had previously relied for their voyages from South America to Europe. It is said that when mounds of tagua ballast piled up in Germany's ports, an artist from Hamburg found that the nuts could be easily carved into figurines, toys and buttons. German businessmen soon saw commercial potential in the nut and by 1895 had set up the first tagua trading post and company in Manta, on Ecuador's coast, Tagua Handelsgesellschaft m.b.H. The tagua crop – or button harvest, as it was known – became an important part of Ecuador's economy, rivalling rubber and cacao as an export.

Buttons, handles and other small objects in the nineteenth century were made chiefly out of elephant ivory, whose trade was famously destructive. 'Every tusk, piece and scrap of ivory in the possession of an Arab trader has been steeped in human blood,' wrote Henry Morton Stanley, the explorer, in 1897. 'Every pound weight has cost the life of a man, woman, or child; for every five pounds a hut has been burned; for every two tusks a whole village has been destroyed; every twenty tusks have been obtained at the price of a district with all its peoples, villages, and plantations. It is simply incredible that, because ivory is required for ornaments or billiard games, the rich heart of Africa should be laid to waste.' I found a beauty in the idea of tagua substituting for this illicit commodity, of a bloodless harvest in the place of distant slaughter. When billiard players struck ivory balls, they might expose bullets or deformities born of violence, but tagua remained untouched by bloodshed.

Thousands of tonnes of nuts passed through the Pacific port of Esmeraldas on their way to the United States and Europe. By the mid-1940s there were in the New York area more than twenty-five factories fashioning tagua into buttons and toys. The largest of these, the Rochester Button Company, employed five hundred people at its peak and produced 3.6 billion buttons a year. 'We may not make the world go round,' the firm asserted in the mid-twentieth century, 'but we button up more garments than any other company, anywhere.' Indeed, the tagua business was so good that in 1948 one of Ecuador's most renowned naturalists, Misael Acosta Solís, warned of the need to protect the nut from over-exploitation: 'It does not matter to the negro or to the "montuvio" [coastal Ecuadorean] whether the forests are destroyed or not; the only thing that interests him is to collect the tagua and get money for his expenses and vices.'

His concern, however, would prove unfounded. During World War Two, thermoplastics substituted natural materials such as tagua and rubber, which could no longer be produced to meet rising demand. As trade in the nuts then plummeted, the Ecuadoreans turned to banana plantations – and, later, to oil – to bolster their economy.

I thought back to the decline of sea silk in the twentieth century in Taranto and Sant'Antioco. It was a gradual decline, a silent disappearance that affected only a handful of women in schools in Sardinia and the Italian mainland. Like sea silk, tagua declined in the post-war era, unable to find a place in the evolving material landscape, but its disappearance was more dramatic, almost tragic, in comparison to that of sea silk. Once the source of countless jobs, an entire culture, this huge export commodity was suddenly relegated to a curiosity, a fringe product, by plastics. In place of tagua palms, the Ecuadoreans encouraged Standard Fruit, the American firm, to clear much of its forests to make way for banana plantations.

The history of materials is largely one of continual replacement: rubber for leather; polyester for wool. Tagua, after all, had replaced other materials, such as horn and mother-of-pearl, and perhaps it was inevitable that its dominance would fade, too. And yet tagua's replacement by plastic was different from previous acts of creative destruction. The word *plastic* derives from the Ancient Greek *plastikos*, which means 'able to be moulded'. Unlike any other natural material, from byssus to the endosperm of a palm, plastic could be shaped to fit our own desires, shedding our attachment to natural structures. 'We divorced ourselves from the materials of the earth, the rock, the wood, the iron ore,' wrote Norman Mailer in the 1960s, 'we looked to new materials which were cooked in vats, long complex derivatives of urine which we called plastic.' There was a permanence to the arrival of plastic, signalling the end of previous

ways. If Western liberal democracy apparently heralded the end of
political history, relegating any alternatives to the fringe, then the
development of plastics heralded the end of material history.

first came across tagua as I wandered through a market in Quito,
Ecuador, during my late teens. I saw carved tagua animals, tagua
jewellery, and, most beautiful of all, creamy-white pebbles of tagua,
pleasing to hold in your hand. Balancing one of these pebbles in my
palm, I felt disoriented: for all its beauty, it must have been extracted
from the flesh of a distant or fantastic animal. Some three million
years ago a land bridge appeared connecting North and South
America, two landmasses that had hitherto remained separate since
the Jurassic era. An array of great beasts headed south: *Hemiauchenia*,
the ancestors of today's guanaco and vicuña; horses; saber-toothed
cats; and gomphotheres, elephant-like giants with four large tusks.
There is little trace of these great creatures today: they had all disappeared by around ten thousand years ago, possibly killed off by
human activity or climate change, but these white seeds, for
me, evoked the gomphotheres that once roamed the land.

When I got home to England, I'd wander through charity shops
trying to find tagua buttons on old suits and clothing from the pre-
plastic age. Slightly faded, perhaps they were harvested in Ecuador
before World War Two, shipped off to a factory in New York state,
and then sewn onto a coat. When writers of fiction imagine time
travel, the transportation of a character to the nineteenth century,
they tend to focus on the sights of the Victorian age: early motorcars;
steam engines; neo-Gothic architecture; revival stained-glass windows. But the world would have felt different, too: the bristles of a
badger against one's skin while shaving; the scrape of a condom

sheath made of treated animal bladder or intestines; the feel of shoe-laces actually made of leather; the smooth contours of a button made from a seed. Touching these tagua buttons, I felt nostalgic for this world that I had never experienced, a time when touch might have connected us to nature.

One day I ordered some tagua buttons online, and they arrived in a small plastic bag. I opened it and played with them in my hands, rubbing their smooth surfaces and inspecting their patterns. Unlike the tagua buttons I had seen on second-hand clothes, these were not faded but creamy and new, fresh off the palm. From the label on the packet, I saw that the buttons in my hand were produced by a firm named Corozo, one of a number of companies that had sought to reintroduce tagua. Drawing on techniques from the nineteenth century, the firm had harvested nuts, cut them to make buttons and marketed them to clothing manufacturers in an attempt to break the monopoly of plastics.

This idea of resurrecting a material from the historical dustheap intrigued me. I took out the vegetable-ivory buttons and put them next to the plastic buttons on my shirtsleeves. How different these buttons were from plastic, whose uniformity defies any attempt to read its history, to fathom its past or origin. Seeds, or their deriva-tives, surround us, as Thor Hanson points out in his book *The Tri-umph of Seeds*: lecithin, from soybeans, is used as an emulsifier and appears in anything from margarine to ceramics; guar gum, from the bean of the guar plant, is used in fracking fluid. Yet there is something different about finding buttons fashioned from raw seeds, their dense cellulose transformed into a fastener. Looking at my tagua buttons, I could see small swirling patterns, which spoke of their origins as a liquid hardening on a palm.

Whenever I held a plastic object, I experienced a light anxiety, tied to the concept of time, the knowledge that it might end up in a distant garbage heap or a raft of floating waste in the ocean. By comparison, these tagua buttons felt light in my hand. Unlike plastic – which may take hundreds of years to decompose, if ever – tagua is biodegradable, its cellulose breaking down on contact with moisture. Like other natural materials, from bamboo to hemp, it offers the possibility of a different relationship with our surroundings: throw away tagua buttons and they will eventually disappear: you need not worry that, by tidying your own home, you are contributing to a garbage pile elsewhere, or that your own true legacy might not be your labour, children or writing, but the thin layer of plastic each of us leaves in our wake.

At the level of production, tagua had something to teach us, too. The plastic buttons on my clothing came from heavy oil, though it is nearly impossible to know from where. Through my own buttons I could be connected to the tar sands of Canada, the Safaniya oil field in the Persian Gulf, or a gusher in Texas. To relate the ecological destruction of oil extraction is almost to repeat a mantra, a predictable history that has been recounted a thousand times, but tagua's history is anything but predictable. Unlike the stem or heart of a palm, tagua can be harvested without damaging the tree to which it belongs. It is a seed, its cellulose serving as a lunchbox for a future palm, but its removal need not spell the end of a line of trees: as the palms produce many nuts in large burrs, it's possible to harvest a portion while allowing for future reproduction. 'It is a product that has such potential for sustainable use and production if it's managed correctly,' Henrik Balslev, a Danish botanist, told me. 'That we like as botanists. We like to demonstrate that nature has something to offer and you can use it without destroying it.'

Henrik's enthusiasm for tagua was just part of a surge in interest among botanists and development experts for the harvest of 'non-timber forest products' (NTFPs), which dates back to the 1990s. If local people could make money from harvesting fruits, seeds, flowers, leaves, roots, bark, latex or resins, it was argued, they might have an incentive to conserve forests. Instead of felling trees for timber, they could forage, prune, clip or crop, much like the eiderdown farmers or vicuña shearers. Inspired by research on NTFPs, Conservation International, a non-governmental organisation, set up a programme in early 1990 to increase the harvest, export and marketing of tagua. According to its figures, over a three-year period in the 1990s, the initiative led to sales of 850 tonnes of tagua in Ecuador and the export of 40 tonnes to the United States and Japan, generating at least $1.5 million in button sales. Firms like Patagonia and Smith & Hawken started putting the buttons on clothing to reduce their use of plastics.

According to the World Wide Fund for Nature (WWF), the greatest cause of species loss is not the illegal wildlife trade (although that is significant), but habitat loss: the encroachment of agriculture, industry, pipelines, housing and roads on forests, lakes, swamps and plains. I saw this in Borneo's caves, which are being quarried for their limestone; in Java, where oil palm has replaced primary forest; and in the puna, where large trucks routinely mow down vicuñas on their way to Cuzco or Lima. It bothered me that none of my objects addressed this fundamental issue: that one could create incentives for protecting vicuñas or civets, but if their habitat, the grassland of the puna or highland forests, were developed, it could threaten their survival. Unlike the other objects, tagua, it seemed, addressed this problem head-on, offering to save more than itself; the palms could house or nourish many species, from insects to deer to squirrels.

As I read about tagua, I became more and more enchanted by its possibilities, my enthusiasm rekindled to eiderdown levels. I looked upon this marble pebble as my own bird's nest, an antidote to all our ills, from deforestation to elephant poaching to the obscene accumulation of plastic waste in landfills and oceans. In fulfilling this purpose, I hoped, it might go beyond itself, undermining the linear conception of our own material history: the maxim that all natural materials would be substituted by the monolith of plastic. Learning from tagua's example, we might return to other materials and, in so doing, regain a sense of history, craft and our place in the world. It was buoyed with this enthusiasm, with tagua on the brain, that I travelled to meet Corozo's founder, an Ecuadorean businessman named Klaus Calderón, in Manta, the one-time heart of the tagua trade.

n the centre of Manta there is a roundabout with a fibreglass model of a huge yellowfin tuna – *Thunnus albacares* – positioned above a replica tin can. Flying above the traffic, the model is about the size of the cars that swirl below. For me it was a display of the advantages of synthetic materials: the monument accurately captures the powerful mass of the tuna, its sheer muscular scale. But it is also a display of the limitations of synthetics, their inability to conjure the beautiful scintillating colours of a tuna, the bright yellow of its fins and the metallic glint of its scales, which flash underwater.

Although its material speaks of glass and plastics, the tuna was intended to serve as a statement of the driving force behind this city. After the post-war decline of tagua, Manta first sought to fashion itself into a commercial port, but it found its true place as a hub for tuna fishing. Since the 1990s, it has been one of the major players in the industry, harbour to most of the country's 116 fishing vessels.

No longer home to small fishing skiffs, its port is crammed with 100- to 1,200-tonne purse seiners, vast combine harvesters of the sea that rely on sonar systems and even helicopters to find fish. Once dumped in the docks, the tuna is transported to the city's processing factories, where it is cooked and compressed into tins up to the size of film reels by the wives or daughters of the fishermen who extract it.

I met Klaus and his wife, Fernanda, in one of Manta's beach hotels, a short walk from the tuna sculpture. Built during the tuna boom, the hotel overlooks one of Manta's better beaches, frequented by Ecuadorean tourists, the occasional American, and long files of brown pelicans that patrol the coastline in search of fish. The day was overcast, the beach sand merging with the sea and sky to form a single murky brown, and it was oppressively hot, even within the air-conditioned hotel. I looked down the menu, which offered an array of seafood, but very little tuna; I later learned that only 20 per cent of the tuna stays in Ecuador, while the rest is exported, mostly to Europe.

Klaus wore white trainers, shorts and a polo shirt and spoke English with an American accent. Sitting opposite him in the hotel bar, one might easily forget that he was Ecuadorean, and mistake him for an American visiting from Miami. Yet his own history is deeply tied to this coastal city and the nut that was once its most important export. In the early twentieth century, Klaus's grandfather left Germany and settled in Manta, taking a job as an administrator at the German Tagua House. A natural storyteller, Klaus delighted in recounting the tales of German and Italian rivalry in the tagua trade.

When Italian button makers first got wind of the trade in the nuts, they sent a young man, Giovanni Zanchi, off to find its source. The Italians' initial suspicion, Klaus said, was that the nuts came from Africa. Zanchi duly headed to Africa to find tagua, spending five years on the continent on a fruitless hunt, until the true location of the tagua supply was revealed by the indiscretions of a drunken sailor. Whatever the truth of this story, the Italians soon set up their own trading house, La Casa Tagua Italiana, across the city's main river, the Tarqui, and eventually came to dominate the nut trade.

It was never Klaus's plan to follow in the footsteps of his grandfather and enter the tagua business. As a university student, he returned home to Guayaquil, Ecuador's largest city, and was introduced to a woman named Fernanda Zanchi, who was working in the same office as his mother. Although he didn't know it at the time, she was the granddaughter of Giovanni Zanchi, the famed nut hunter. 'By chance I met Fernanda and then we went steady for a while,' Klaus told me. 'One day [the Zanchis] decided to invite my mother to their home in Manta, and then sitting after lunch . . . we came to talk about family history. We were very much surprised [by the connection]. It's unbelievable.' Klaus and Fernanda got married shortly after meeting. 'It was impossible to share such a common past and not feel attached to one another,' he said.

In 1979 they established their own business, Corozo Buttons, selling blanks – unfinished buttons – to button producers in Italy. It was a good time to introduce tagua to the market, given the increasing awareness of the drawbacks of our wholesale reliance on plastics, and their customer base gradually expanded. When Klaus spoke about tagua, he became animated, enthusing about its textures and swirls; at the mention of plastic buttons, he recoiled in disgust. He'd scoured historical sources to reconstruct the history of this nut, considered tagua from every angle, and explored its potential uses, even as an energy drink in its liquid state. I found his enthusiasm contagious. All around us in the hotel we were surrounded by plastic:

polyester tablecloths, polypropylene chairs, and my own buttons, all signs of the victory of synthetics. He wanted to counteract that narrative, introducing this nut that would revive this family industry.

At the end of the eighteenth century, two Spanish botanists, Hipólito Ruiz López and José Antonio Pavón Jiménez, headed to South America to gather specimens. Travelling through the eastern foothills of the Andes, they came across tagua palms and named them *Phytelephas*, combining the Greek words for 'plant' and 'elephant'. 'They reasoned', explains the Danish botanist Anders Barfod, 'that species of *Phytelephas* were ivory bearing plants just as the elephant is an ivory bearing animal.' The botanical family of Arecaceae, or palm trees, spans some 2,500 species. Mostly confined to tropical or subtropical climes, they produce an array of seeds of different shapes, textures and sizes, from the coconut to date stones, but for me nothing could outstrip the strangeness of tagua, the elephant plant.

I sought to find the source of the ivory, not to exploit it but to feel it, to place it within a tree. I got the name of a tagua trader from a source in Manta and took a bus to meet him in the town where he lived, Jipijapa, some fifty kilometres inland. The change in altitude was measured by the thickness of the trees: the large swollen ceibas of Ecuador's dry coast were gradually replaced by *guayacán* and acacia trees. When I got to Jipijapa, the dealer gave me a couple more names of tagua dealers in a village called Pedro Pablo Gómez; they, in turn, told me to head to La Crucita, a small collection of wooden houses in the forest. My journey had a hint of absurdity to it; one name simply led to another, and I began to fear that the tagua supply chain was a labyrinth or a circle, leading back on itself, but by

late afternoon I was relieved to arrive at its starting point, meeting
Franklin Pilay, a campesino who harvested tagua.

A muscled and stocky man in thick rubber boots, Franklin lived
in a wooden cabin built by his father, which he shared with his family. Probably in his sixties, he spoke little and quietly, expressing his
emotions through laughter and gestures. He greeted me with great
openness and generosity, possibly associating me with USAID staff
who had worked in the village, training locals to carve tagua figurines for the tourist market.

His *tagual*, or tagua grove, lay a short walk from his home, down
a dirt track lined by avocado and guanabana trees, whose large
sweet fruit reminded me of the durians I had seen in Malaysia. Following him along the track, I passed a field with several tagua palms,
rising high above the grass like oaks in the English countryside. I
admired the large spadices of the male tagua palms, huge yellow
tongues that drooped downward, wafting their pollen to the female
trees with their smaller inflorescences. By the side of the road,
we shortly came across an uprooted tagua palm that still had its
large burrs the size of footballs. With a power that belied his short
stature, Franklin hacked at one of the heads with his machete, cutting a slice off its top to reveal five unripened white seeds. Still soft,
they looked like boiled eggs, bright white against the brown fibrous
walls of the burrs.

It's said that Henry II's knights sliced off the crown of Thomas
Becket's head at Canterbury, 'in such a way that the blood white
with the brain, and the brain no less red from the blood'. Such was
the intensity of the colour of these seeds that I felt that we had cut into
something human, revealing its brains, bones or teeth. My first reaction was to recoil in disgust, but when Franklin pulled out one of
the seeds from the casing, I felt only delight, rolling the white ball

between my own palms, its smoothness contrasting with the fibrous shell of the burr. At Franklin's suggestion, I bit into the endosperm, but my teeth only scraped the seed's surface, gouging small grooves. Sweet and milky, it tasted like coconut, a flesh that I crave, but I had little inclination to break my own teeth against the vegetable ivory of the elephant tree. Too hard to be eaten, but not solid enough to make buttons, this seed could give pleasure only as an object, telling the story of tagua's transformation from liquid to nut. I slipped it into my pocket, another addition to the cabinet, and pressed it between my fingers and palm.

The words for *palm*, the tree, and *palm*, of our hand, share a common root, tied to the concept of spreading. When we reached Franklin's own land, I could sense the aptness of this connection. On a steep slope I saw at least twenty tagua palms, their fronds spreading downwards, huge feathered fingers that brushed us as we passed. He pointed out males and females, teaching me to spot the differences between the two sexes. Unlike anyone I met in Manta, Franklin did not call them tagua, but *cade* or *cadi*, a Quechua word whose meaning, 'thatch', evokes the place of these palms in the lives of those who dwell in their vicinity.

At the base of one of the palms, Franklin spotted a small tagua seedling, reaching no higher than my boot. He grabbed its miniature leaves in a single clump and cleared away the vegetation that surrounded it with his machete. It would, Franklin said, be at least ten years before it bore any fruit. Unlike oil palms, tagua palms are ill suited to agriculture; their seeds may remain dormant for long periods, and they grow incredibly slowly. In 1948, close to the era of tagua's boom, Misael Acosta Solís, the Ecuadorean biologist, fretted that tagua seeds would be smuggled out of Ecuador and planted elsewhere, much like cinchona and rubber. Yet his worries proved unnecessary: unlike cinchona or rubber, no tagua seeds were ever smuggled out of South America to start large plantations, even at the height of the tagua boom; the palm was just too impractical for large-scale cultivation, predestined to be foraged wild.

We rummaged around the bases of female trees, frightening the squirrels that were feasting on the orange mesocarp of the fallen burrs. Foraging in the popular imagination is often associated with mushrooming trips, a family affair, but Franklin's hunt for tagua was anything but romantic; lancehead vipers – *Bothrops atrox* – often frequent the bases of palms, and it's not unknown for tagua foragers to be incapacitated or even killed by their venom. That day we found neither snakes nor nuts, only the remnants of the burrs, fibre, and flesh. Frustrated, Franklin tried another approach. As agile as a teenager, he climbed the trunk of a female elephant tree in his rubber boots, his powerful hands gripping old leaf bases. Hanging on to the leaves, he inspected the large black heads and tapped their fibrous walls with his machete. None were ripe, he said, their outer surfaces still hard.

Watching him scale the tree, I thought of the birds' nest harvesters climbing the poles, risking their limbs to scrape swiftlet saliva. And yet tagua, I reflected, is not a nest, but more akin to a fertilised egg. When you harvest tagua seeds from a tree or the forest floor, you are removing not a home, a form of protection, but the potential for regeneration and dispersal. If carried out thoughtlessly, tagua harvesting can often be destructive, reducing the chance that palms may reproduce, but here there seemed little risk of that possibility. Conscious of the health of his palms, Franklin did not hack the unripe burrs he had found, preferring to wait until they softened. Disappointed by their hardness, he walked down the trunk, clinging to a leaf for support, and staggered back down to the dirt track. I had almost lost hope that we would ever find some seeds, but on our descent, Franklin spotted a cracked burr high up that had not yet released its entire cargo. He promptly fashioned a pole from a branch and prodded the orange flesh, disgorging the seeds, which rained down in their orange pouches. Overjoyed, he let out a yelp of pleasure.

Since the initial enthusiasm for NTFPs in the 1990s, researchers currently treat the merits of commodification of forests with greater caution. It has been argued that the harvest of a particular fruit or seed is no guarantee of the survival of other species. Certainly this has been proven in the case of tagua, where most trees are harvested in stands that have been cleared of other vegetation. Franklin's own land was no exception. Weeded of other species, it was clearly a landscape subject to intense management, more orchard than dense primary forest.

I visualised all the other landscapes I had seen, from the puna to the Westfjords to the swiftlet cities. Weeded of unprofitable species, from pumas to guanacos to mossy-nest swiftlets (which build their nests from grass), were they any different from these tagua stands? All my objects offered the prospect of salvation, yes, but it was clearly a limited salvation, one predicated upon our own desires. Even here it was not clear that tagua would continue to be part of the land-scapes we created. On the return from Franklin's land, I noted several blackened palms, charred as if they'd been struck by light-ning. Treated as weeds, they had been burned by another farmer, as the seeds' price, seven to nine dollars per quintal (100 pounds), was not high enough to make a living. Even Franklin rarely har-vested tagua, relying on other crops like corn; his cutting machines lay unused, rusted after the departure of American development specialists.

Such thoughts did not interrupt my delight with the treasure we had found, the six or seven seeds that I jealously cradled in my arms. I didn't care that tagua could not save the forest, that the value of these seeds, which had taken us hours to find, was only a few cents: they were beautiful and at that moment this was enough for me. I'd read all about the history of elephant hunting, the bloody tales of Victorian adventurers or Theodore Roosevelt. Then I'd watched footage of John Hume's staff tranquillising one of his rhinos with a dart, before removing its horns with an electric saw, leaving two stumps. I'd watched it stagger back up, returning to life in its enclosure. I now held ivory without any of this, touching this texture merely by gathering a seed. We went back to Franklin's home and handed the nuts to his nephew and nieces, who devoured the orange flesh to reveal the brown nuts, the first part of the button production line.

f one were able to trace the history of a plastic button, it is not likely that it would start in Rochester, or indeed anywhere in New York state. After decades of decline, the Rochester Button Company, once the largest producer of buttons in the United States, reportedly closed its doors for good in 1990. Instead the journey would begin in a town in southern Zhejiang province, China, named Qiaotou. Qiaotou is just one so-called industrial cluster in Zhejiang: the area includes a host of others dedicated to all manner of products, from lighters to glasses, shoes, pumps, valves and pens. Prior to Deng's free-market revolution, it was a farming village, but in the late 1970s two enterprising brothers, Ye Kelin and Ye Kuchun, started selling buttons to local manufacturers; by the 1980s, Qiaotou had started manufacturing, producing cheap plastic buttons for Italian clothing firms. Now they supply an estimated 60 per cent of the world's buttons.

Although Qiaotou makes buttons from many different materials, the majority are made from plastic, itself derived from cheap heavy oil. Hundreds of factories mould the plastic into buttons or cut them from plastic sheeting, churning out thousands every minute. By contrast, the process of making tagua buttons in Manta is more Victorian than high-tech, a far cry from Qiaotou. Stripped of the orange flesh, Franklin's nuts, which resemble avocado pits, would head to Manta, via the endless pairs of hands I had shaken on my way to La Crucita. After drying out in the sun, they would be rasped, scraped, sliced and punched into small circles – blanks – in modest workshops throughout the city, before being exported to button makers in different countries. 'If you just drive by one of these neighbourhoods [in Manta]', Klaus told me, 'you open the windows, you start listening to the saws cutting and the other machines turning.'

I did as Klaus suggested, taking a taxi through Manta with the windows down, crossing the river that once divided the Italian and German operations. I didn't hear any cutting of tagua and sensed only the overpowering smell of the fish-processing factories, a reek stronger than that of Nuar's kingdom. Instead of the German and Italian tagua houses, I spotted the monument to the tuna trade, strip malls and large high-rises overlooking the beach, testimony to the cash from the city's tuna fleets and, some say, the cocaine that is often smuggled in their hulls. Crossing the river, I came to the poorer side of the city, much of which would later be destroyed during the 2016 earthquake that hit Ecuador's coast. Rubbish lay strewn across the riverbed and, oddly, a sofa lay in the middle of the beach, facing toward the surf. Much less touristy than Murciélago (Bat) Beach on the southern side of the river, Tarqui Beach is frequented by local fishermen hauling in their catch of marlin and shark, which line the sand in bloody files.

Unable to find any workshops by ear, I got in touch with Maritza Cárdenas, a local button producer, who is head of Fedetagua, an association of small producers. She offered to take me to a few workshops to see the cutting of tagua. We went into a dark workshop in Manta, deafened by the roar of saws, music and the squeals of pigs being slaughtered next door. Seated at individual workstations, men and women were busy slicing tagua into disks with metal saws. Within the room there were only five people cutting, but Maritza said that each one supported a large family, kept afloat by the nut. 'This is one family,' said Maritza, gesturing toward one worker at a single unit. 'Only in this one there are ten people.'

I'd read accounts about the difficulties of working with leather, this natural material, in my great-grandfather's tannery in Runcorn. It was hard, insalubrious work, requiring labourers to dip hides into pits full of lime to loosen the hair and plump the skin, before they were placed in tannic acid, left to soak, and then dried. Tagua was different from leather; it didn't require huge physical strength to cut it, but the work was similarly demanding and dangerous: the tagua cutters tended to wear metal finger guards to protect themselves from the saws, mufflers over their ears, and bandanas over their mouths to block out the dust from the nuts.

In the corners of workshops, I would often find bags and bags of what appeared to be ground flour. Unlike plastics, tagua cannot be moulded – a simple fact that takes up workers' time and results in large quantities of waste. Once the blanks are punched from the nut, the workers are left with about 85 to 95 per cent of the original seed, which is normally then ground up into a white powder often used as livestock feed or as a combustible for the tuna-processing factories. The amount of waste so perturbed Klaus that he tried pouring the liquid endosperm of tagua into button-shaped moulds, but so far his experiments had not borne fruit, appearing almost alchemical in their aspirations.

I took one of the tagua disks and rubbed my fingers over its smooth creamy surface, the solidified endosperm of the palm. I had fallen for the idea of tagua, this connection with a palm through a button, but I also saw that the reversion to this natural mould was a reversion to certain labour practices. My ears deafened by the sound of saws, my eyes glazed by the repetitive motion of the cutters, I found little that was romantic here, only the endless production of disks and blanks, infinite cream circles like miniature backgammon pieces. In 2010, Ai Weiwei brought his exhibit *Sunflower Seeds* to the Tate Modern in London. I remember visiting it as a graduate student, peering down at what appeared to be one hundred million sunflower seeds laid out in the Turbine Hall. From a distance these seeds appear to be natural, but up close I realised that they were made of ceramic, handcrafted by hundreds of workers at Jingdezhen, a city in China renowned for its porcelain production. I felt a combination of awe and horror, struggling to imagine the number of hands and the amount of time required to produce such a work. Surveying the infinite button blanks in Manta, I experienced a similar emotion.

There are perhaps hundreds of these small workshops in Manta, churning out blanks, but their sheer number disguises the fact that tagua producers in Ecuador are in a state of constant crisis. Unable to compete with China or local informal production, Manta's tagua button factories, including the one run by Klaus, closed in the

1990s, sending their workers into informal cooperatives, while the factories' one-time owners became exporters. Even some of the exporters have now given up; dismayed by the market, Klaus headed to Panama to set up a new business, this time producing finished tagua buttons, using different dyes and designs. The workers, the actual tagua cutters, evoked not epic stories of exploration or German and Italian rivalry, but the strains of producing high volumes of buttons by hand for low market prices. Unable to make ends meet, many had left to work in the tuna-processing factories or hoped to find employment in an oil refinery, which the government had pledged to build outside Manta with Chinese financing.

Explaining the crisis, some blamed the exporters, the middlemen who had access to markets, the increased regulation and taxation under President Rafael Correa, or trends in the global market. But the main issue is the material itself, the very nature of this seed. Not only are they expensive to produce, but tagua buttons need to be treated with care; prone to swell during washings, they may cut their threads if they are tightly sewn onto garments. Shortly after starting to sell clothing with tagua buttons in the early 1990s, Patagonia received reports of breakages and replaced more than a quarter of a million tagua buttons. It was a bold idea to use tagua, but it ended up inflicting 'a significant cost in dollars' on the company. The world did not want tagua, did not want its inconvenience, its biodegradability, preferring the plasticity of plastic, this eternal material. When it came down to it, I was no different; when I looked at my own body, I was a plastic man, a walking vessel for plastic from the soles of my feet to the resin in my fillings.

Only a single tagua export business seemed to be thriving, one that attracted the ire of many tagua cutters. It was rarely mentioned by its real name; the cutters simply called it 'the Big One'. The Big One, I later learned, was Trafino SA, Ecuador's largest tagua exporter, which now controls some 80 per cent of the market. It buys blanks directly from small producers, and sells them to button makers in China, Turkey, Germany, Italy and Spain, who in turn sell them to major fashion houses. The owner of Trafino is Francisco Luna, a short, stubbled man with a surprisingly deep voice. He lives in a glass tower block at Barbasquillo beach, almost indistinguishable from the hotels that lie adjacent to it. Far from the smell of the tuna factories and tagua workshops, his building boasts its own Jacuzzi, swimming pool, and private beach. 'This city has a bad smell,' he told me as he greeted me in the marble lobby.

Francisco had recently produced a documentary about tagua. Like Harrisson's film, it journeys from raw material to finished product, starting in Ecuador's forest and ending in button factories in Turkey and China. It presents tagua in near-heroic terms, a material that can easily be carved, a source of income for campesinos, and a tool in the fight against deforestation. In one breathtaking drone shot, we see aerial footage of a rich canopy, chaotic and beautiful, slowly replaced by the ordered spacings of an oil palm plantation, the different greens and textures of the forest dissolving into the single hue of a foreign palm. Not enough people, it suggests, are buying tagua; no one has any incentive to protect the palms. 'A collector goes out with his donkeys and gets two quintals that go for three dollars each. With six dollars he cannot even eat,' Francisco said. 'The only way to solve this is to have a decent price.'

Unlike many people whom I met in Manta, Francisco did not have a family connection to tagua, or ties to the Italian or German export houses that flourished in the early twentieth century. An engineer by training, he'd started out building aluminium structures in Quito and Guayaquil, specialising in covers for a stadium in each city. 'These two works set world records', he said. 'And I have them both!' After falling out with his business partner in the late 1980s, he came to Manta, where he set up his own tagua factory, producing blanks. It flourished initially, he said, employing some sixty people, but eventually foundered in the face of competition from small-scale producers. 'As it was very easy to produce the product, many informal factories sprang up in homes,' he said. 'They put up four canes, a roof, stole the electricity from the grid, and started working.' He closed shop and started exporting instead. 'We export five times as much as the second-largest exporter,' he said proudly.

While I was in Manta I met enthusiasts who worked in tagua, obsessed by its history and charm, accepting its limited application. Yet Francisco was somehow different. While I was waiting for him in his apartment lobby, I heard him talk with a business associate in a tone that was almost a caricature of the hard-nosed businessman. 'I want results!' he bellowed, shortly before ending the call. Drawn to scale, he spoke like a construction magnate rather than a nut en-thusiast, dwelling on the square metres of his aluminium covers, the percentage control of the tagua market held by Trafino and his po-litical connections in Ecuador's capital. I asked him about the com-ments of some of the small-scale tagua producers, who claimed that he was running them into the ground by buying at such low prices. As if talking to a child, Francisco calmly rebutted their arguments, citing changes in global prices. 'This year demand went down – so prices go down,' he said coolly. 'This is the market. There is nothing that can be done against the law of the market. The law of supply and demand is stronger than God.'

Tagua, for me, has always been associated with charm ever since I started collecting it in my teenage years, its varied forms populating my shelves. I would gift tagua objects – a small pipe, thimble, or salt shaker – to good friends, telling them stories about its history. I'd give back rubs to family with the edges of tagua nuts, explaining why they are so smooth. I wanted to believe in its promise, its ability to resolve difficult problems, but the more time I spent in Manta, the more I became aware of its limitations. Like so many materials we vaunt and idolise, this nut – this solidified endosperm – could not save elephants, significantly reduce the use of plastics, curb deforestation or even provide much of a decent standard of living for many of the people who cut and punched it. Even if Francisco Luna got what he wanted, and the demand for tagua spiked, I found it hard to believe that those gains would eventually trickle down to those who harvest the nuts; they would remain captured, coagulated in the pipeline, by exporters, as in every other commodity chain in existence.

'It's sad when you have been there and seen the harvesters,' Henrik Balslev told me. 'They invented it all and did the hard work. They started the whole thing and end up with nothing.' I thought of the eiderdown harvesters, swiftlet nest collectors, civet coffee farmers and vicuña shearers, these disparate groups at the bottom of these trades. Despite the differing textures or functions of down, saliva and seeds, the human relationships within each chain remained the same, a site of constant conflict. You could take the objects, arrange them in different orders, but the patterns they formed were always the same: to follow each object was to trace the flow of a liquid that, having welled up like oil at the harvest, flowed outward in channels, coagulating in distant pools, from offshore accounts to country houses. 'For every producer I have known it's been like that,' Henrik said.

More than disillusion with a nut, with its narrative possibilities,

I felt the loss of an idea that an object, a material, could resolve certain problems. Frustrated, I spent more and more time in the small apartment I had rented in Manta, watching brown pelicans patrol the beach in their long files. I went out little, losing the sort of manic extroversion that one needs to follow objects. I spent time watching films, talking with friends and botanists on Skype, supposedly doing research, but really avoiding the city, whose smell and heat now repulsed me. On the rare occasions when I went out, I sometimes saw two American Jehovah's Witnesses, a young couple in immaculately pressed clothing, pulling a suitcase full of reading material behind them. I felt strangely envious of their poise and confidence without really knowing why and grew careless and apathetic. As I stripped back the layers of tagua burrs, ripping off their fibrous casing, their orange flesh and brown skin, I became more and more melancholic: all that remained was the beauty of this white pebble, this palm from the elephant tree within my own.

GUANO

n a Victorian country house close to Bristol, not far from where I grew up, I hunted for traces of eiderdown through its labyrinthine corridors and bedrooms. After I followed six other objects, the search for this material had become more than an idle hobby; it was something of an imperative. As civet coffee, sea silk and now tagua failed to live up to their promises, their histories resisting idealization or templates, I found it reassuring to think that this material still offered the possibility of balance, for all of its bloodiness and compromise. I had read in the National Trust catalogue that the house, Tyntesfield, had an array of eiderdown items, covered in silk. Yet when I entered the bedrooms, I found that these duvets lacked eiderdown's volume, its lightness: they were simply stuffed with feathers from dead ducks or geese.

Reading up on *eiderdown*'s etymology, I found a surprise: sometime in the nineteenth century the very word became untethered from the duck; far from the down of an eider, it came to mean a duvet with any filling whatsoever or a fabric of thick texture. In a UK parliamentary debate in the 1970s, the Labour politician Stan Orme mocked MPs in the House of Commons, explaining that 'they have joined the eiderdown gang'. He meant, of course, not that his political opponents had a craving for duck down, but simply that they had been in bed rather than discussing the intricacies of trade union legislation. I was disconcerted to learn of eiderdown's expansion, not least because it suggested that all those eiderdown duvets I had read about in travellers' accounts may have had nothing to do with the sea duck. Like all those writers working on sea silk, I may have inadvertently added to errors, assumptions, myths and falsehoods.

The eiderdown I had found could perhaps stand as a symbol for the house I found myself in. Tyntesfield is at once a monument to the extraordinary wealth of its owner and an attempt to reconcile it with his Anglo-Catholic piety. Its collection of objects could compete with any of the great collections of the seventeenth century: when I visited, losing myself in its countless rooms, I found bone china, 'eiderdowns', and a large billiard table with ivory balls. A member of the family who once owned the house was a keen ivory turner, carving hundreds of small ornate objects, whose beauty, I did not like to admit, far outstripped any tagua object I had seen. Yet I found the house's surfaces, its wood panels, silk curtains and celestial symbols, unable to evoke the earthly materials from which it was built.

Unlike Ole Worm's museum, Tyntesfield is not a display of curiosity, learning or intellect, but a monument to capital. Its owner, William Gibbs, once headed one of the great trading houses of the Victorian era, Antony Gibbs & Sons. William's father, Antony, had founded the firm, dealing in a whole range of commodities, but Gibbs *fils* ended up specialising in a single material, a move that would bring him fantastic wealth. Roaming through the house, I found little sign of what this material might be. The only clue to the riddle of its origin was a design on a stained-glass window. Upon it were birds striking different poses, preening themselves or standing erect. Unlike those of swiftlets, their feet were not minute claws, but cumbersome, akin to those of waders or seabirds. Their beaks, sturdy and thick, resembled small daggers, tools for fishing rather than for catching insects.

I read a degree of reverence in the position of these birds, overlooking visitors as if they were objects of adoration, much like the religious symbols found in Tyntesfield's chapel. It is possible that the design on that window represents Peruvian seabirds, the actual source of Gibbs's tremendous wealth. In the nineteenth century they nested in huge colonies on some eighty islands off the coast of Peru, nourished by abundant fish in its plankton-rich waters. Over centuries their excrement had accumulated on these islands, a rich source of phosphates and nitrogen, key ingredients for crop growth. In most places around the world, bird excrement is leached by the presence of moisture, but these islands off the coast of Peru receive little or no rainfall. Baked in the hot sun, the excrement, known as guano, had simply accumulated, a vast source of organic fertiliser.

In 1842, Gibbs's representatives in Lima took out a contract with the Peruvian government, providing financing in return for rights to market the guano. At the time, Gibbs was unsure of the soundness of the decision, but his firm's gamble proved astute. All over the world, particularly in Europe and North America, farmers and plantation owners complained that their fields were exhausted because of the demands of growing urban populations. Offering a chemical balance that outclassed its competitors, guano came to provide the solution. By the 1850s guano had become the single most important fertiliser in the world. According to the American historian Gregory T. Cushman, over a period of forty years Peru exported an estimated 12.7 million tonnes of guano from its islands, to destinations all over the globe. 'The sunless lands of Great Britain, the rice fields of Italy, the vine lands of Germany, the exhausted coffee plantations of Brazil, and the arid plains of Peru, all testify to its fertilising properties,' enthused *The Farmer's Magazine* in 1855.

The success of Peruvian guano set off a global rush for other sources of excrement. Desperate for their own supplies, adventurers and businessmen explored every scrap of land, atoll and key in an attempt to find guano, from Scottish crags to islands off the coast of South West Africa. In 1856, at the urging of American farmers, the United States Congress passed the Guano Islands Act, authorising any American citizen to seize an island with guano deposits; US citizens later claimed sixty-six islands in the Caribbean, Oceania and the Pacific. But none of these finds could compare with the deposits of rainless islands. 'I don't think much of the new discoveries,' wrote Henry Gibbs, William's nephew, in 1855. 'I hear of none where the rain does not fall – and as long as *their* Rain continues our Reign will also. So we will sing "Long to rain over them!"'

The guano trade generated colossal wealth for both the Gibbses and the Peruvian government, which owned the deposits. 'So great is the value of this branch of national riches, that without exaggeration it may be affirmed that on its estimation and good handling depend the subsistence of the State, the maintenance of its credit, the future of its increase, and the preservation of public order,' wrote Manuel Ortiz de Zevallos, Peru's finance minister, in 1858. Perched at the top of the trade, William Gibbs became wildly rich. Between 1842 and 1875, Antony Gibbs & Sons earned up to £100,000 a year, of which William received between 50 and almost 70 per cent from 1847. It is said that the following rhyme was heard in London: 'The House of Gibbs that made their dibs / By selling turds of foreign birds.'

That a fortune could be built on birds should perhaps not surprise us. Consider Dr Boedi's house in a gated community in

Jakarta stuffed full of treasures; or the stock value of food firms, such as Tyson Foods or Pilgrim's, that slaughter millions of intensively farmed chickens each week. The strangeness of Tyntesfield is not its relationship to boobies, cormorants or pelicans, but the nature of their derivative: that this Victorian Gothic pile was built not of bird flesh, feathers, nests or eggs, but of excrement. Wandering through its corridors, I reflected on this supreme and hidden act of alchemy, this conversion of shit into turrets, stained glass and a chapel. How did this act of alchemy take place? Who performed it, and what was their relationship with the birds that laid the golden eggs?

So many of the objects in this book perform a necessary function in the natural world. Byssus serves as an anchor to a mollusc, tethering it to the sea floor in the face of currents; the fibre of a vicuña protects it from both the rays of the sun and the extreme cold of the puna; the endosperm of a palm provides the seed with valuable nutrition. Yet excrement is a waste product, one that has been rejected by an organism. Remove it, I imagined, and there was little need to counterbalance that action, to replace it with a substitute or indeed limit one's harvest. By virtue of its very nature, guano proffered the possibility of a different relationship, one unburdened by any obligations; a harvester could take what he or she wanted without any risk of destruction.

However beautiful I found Tyntesfield, it shed little light on the precise method of the guano harvest, its system of rights and obligations. Aside from the birds on the window, I found little hint of the guano trade. Rather than fish or ammonia, the house smelled of dust, cleaning fluid or the perfume or odour of the thousands of visitors who come each year. It was the domestic life of the Gibbs family, rather than the habits of boobies, that dominated the house. One could learn about the tastes of the family, the predilection of

William's son Antony for ivory turning, or the design of the family's aviary, but little about the extraction of guano. Those great piles of excrement thousands of miles away in the Pacific seemed unconnected with this Victorian pile, which had apparently emerged from the English countryside in an act of spontaneous generation.

One summer I travelled to the Chincha Islands, some twenty kilometres off the coast of Peru. At three a.m. in the Peruvian port city of Pisco, laden with fresh fruit and water, I found a fisherman to take me out to the islands. Unused to taking passengers, he expressed surprise that I should want to go there. 'There is nothing there,' he said. It was dark, the port lit only by the lightbulbs from the fishing boats, but I could make out that he wore a thick grey woollen jumper stained with octopus ink. After a journey of three hours, accompanied by sea lions and great lines of guanay cormorants, I could make out a set of three small specks of land, though their presence had been signalled earlier by the reek of guano, a smell that somehow combined the sharpness of cleaning fluids and the warm odour of a fish market near the end of a hot day.

When seafarers visited the Chinchas in the nineteenth century, they found great mounds of excrement, some eleven stories high. Like sedimentary rock, guano compacts under its own weight to form dark strata the consistency of soap, burying itself and anything it touches: trinkets, artefacts and at least one mummified penguin have been found within its layers. Approaching the islands, I saw that little was left of those great heaps, their presence evidenced only by nineteenth-century reports of surveyors or rare black-and-white photographs. Instead the islands were encased in a thin white layer of guano, masking the dark granite of which they are made;

only at the skirt of the land was its true nature discernible, a fleshy redness revealed by the gentle lapping of guano-milky water.

Seasick and exhausted, I clambered up a rope ladder towards the hanging dock of one of the islands, my progress hampered by the fresh fruit I had brought. I was greeted by a stocky man with several gold teeth. This was Flaviano, the island's guard, who lived in a soggy wooden cabin. He mumbled something, took my access permit, and walked to his office, a small room facing out towards the water. As he filled out some paperwork, I noted a Peruvian flag, a map of bird nesting sites, a calendar from a toothpaste manufacturer, and a cut-out of a woman from an adult magazine glued to a haggard clip-board, the first of many such clippings I would later find on the island. Flaviano asked me to sign the visitors' book and I wrote my name on the first line. 'Welcome to Chincha Norte,' he said, his voice almost inaudible over the roar of birds.

If the guano trade ended in country houses or in food to be eaten by city dwellers, then its source, its point of origin, was the Chincha Islands. Once home to the largest guano deposits in the world, they were believed to hold enough excrement to supply farmers with cheap fertiliser for a thousand years. Throughout the 1840s and 1850s, hundreds upon hundreds of ships headed to the Chinchas and Peru's other guano islands to bear off their riches. Like coal, the mineral with which guano was often compared, their deposits were surveyed, mined and sold. Those who dug the guano saw it as separate from the birds, chasing them away as nuisances or feasting on their flesh. Within decades the bird population on these islands had collapsed, an unrivalled ecological disaster.

I thought of the ransacking of Niah, the conversion of the Great Cave into a mine, but the Chinchas were more extreme sites of extraction and cruelty. Given the insalubrious conditions on these islands, few free men were willing to work here; from 1849 onwards, a Peruvian businessman, Domingo Elías, resorted to importing 'coolies', indentured labourers from China, to hack away the guano. By 1853, it was reported that there were eight hundred Chinese on the Chinchas alone, toiling in the equatorial sun six or seven days a week to work off their debt. According to eyewitness reports, floggings were commonplace, and many died from overwork or suicide; visitors remarked that corpses littered the islands. 'That they are worked to death,' wrote George Washington Peck, an American traveller, 'is as apparent as that the hack horses in our cities are used up in the same manner.'

In the legal sense the indentured labourers were not slaves, but the guano islands dissolved the distinction between the two. 'Once on the islands', wrote one eyewitness, '[a worker] seldom gets off, but remains a slave, to die there.' According to the historian Watt Stewart, the labourers' decision to leave China was prompted by the political instability of the Taiping Civil War. More than ten thousand miles from Peru, they had little understanding of the contracts they signed, and no inkling of where they would be heading; coercion, deception, kidnapping and crimping were all commonplace. It is estimated that over twenty-seven years, some ninety-two thousand Chinese indentured labourers started the journey to Peru's plantations and guano islands, and of these 10 per cent perished en route.

Ensconced in Lima or Callao, the merchants who traded guano were largely distanced from the nature of its extraction, interpreting the material through ledgers and figures. There was little concern with the digging, lugging and attrition of guano extraction. In Europe guano was advertised as a life-giving substance. 'If ever a philosopher's stone, the elixir of life, the infallible catholican, the universal solvent, or the perpetual motion were discovered', reported *The Farmer's Magazine* in 1854, 'it is the application of guano in agriculture.' Yet the closer one got to the islands, to the centre of its production, the more it appeared that guano's true meaning was not life, but death. 'The islands', wrote Peck, 'seem to me to be a kind of human abattoir, or slaughter-house of men.'

As Flaviano and I walked around the island, it was hard to reconstruct this history, to visualise this slaughterhouse. Lizards scattered away from our feet, almost gliding across the guano. Where this boomtown once lay, the cane huts of the diggers, all I saw were the densely packed craters of the nests of boobies and cormorants. Where hundreds of ship masts had once formed a forest in the island's coves and bays, there were only a handful of small fishing skiffs. I asked Flaviano whether he knew of any trace of the past, and he offered to show me the *'cementerio'* on the other side of the island. The sun had now risen, its rays reflecting off the guano like snow, burning the undersides of my chin and nose. Crunching the guano in my leather boots, I found it hard to orient myself: there were no trees, few fixed points on this river of excrement. Guano blind, I followed Flaviano closely, dependent on him as if he were a mountain guide in this glacial wasteland. After some fifteen minutes, we came to what he had wanted to show me. It was a small iron cross, rusted and wonky, fashioned from two bits of metal. 'Here it is,' he said grandly. 'This is the Chinese cemetery.'

I looked around, examining the field of excrement that seemed to stretch interminably into the distance. There was a patch where the guano seemed a bit darker, as if scorched by fire. Yet before us there were no headstones, no markers aside from this small cross, nor any human remains. Deprived of a resting place, the remains of the indentured labourers had been burned or dug up with the guano, then packed up and shipped off; the bones of men and birds had combined, fertiliser for a foreign field. On my return from the *cementerio*, I decided to go for a swim, making my way past Flaviano's cabin and the latrine suspended over the cliff and on down towards the beach. Walking on the sand, I noticed a young sea lion, which seemed not to fear my presence. Approaching it, I saw that it was still alive, but that its insides had been opened up by the beak of a turkey vulture to reveal a bright red against the yellows and whites of the guano and the light brown of fur. This sudden burst of colour some-how brought this history to life, adding colour to the black-and-white photographs I had with me. I did not swim that day.

If the first object in this book stood for a utopian vision, then the guano of the nineteenth century stood for its very opposite, a counter-eiderdown. When I think back to the other objects, I could not find a site, not the caves of Niah or the cages of Bali, that matched the frenzied nature of the extraction on the Chinchas. When Alexander Duffield, a British engineer, visited the islands in the 1870s, he reported that 'these same islands looked like creatures whose heads had been cut off, or like vast sarcophagi, like anything in short that reminds one of death and the grave.' By 1870 the depos-its had been totally exhausted; Gibbs shrewdly moved on, marketing Chilean nitrates instead.

In the early 1860s an American photographer, Henry Moulton, visited the Chincha Islands. Shortly before his arrival, the Spanish had occupied the islands, sending a batch of marines to seize the guano and paralyse the Peruvian state. It is a significant moment in history, the last-ditch attempt by the Spanish to restore their in-fluence in South America after the loss of their colonies, but it was to the guano deposits that Moulton paid most attention, their various forms, shadows and textures. Look closely at Moulton's photographs and you see a mountain under assault, its sides fluted by pickaxes. Small dots at the base of the mound, the diggers stand where the guano once rose, eating away at its flank as flies consume flesh.

As the Great Heap was hacked up, William Gibbs embarked on an extensive renovation of Tyntesfield, employing the services of John Norton, a renowned Gothic Revival architect. He doubled the house's size, adding towers, gables, tourelles and crenellations to the roof, multiplying its stained-glass windows, sheets upon sheets of lead glass. After visiting Rome, Gibbs commissioned Lawrence Macdonald to sculpt a white marble bust of his daughter cradling a small bird. In 1872, Gibbs asked Arthur Blomfield to add a chapel – modelled on Sainte-Chapelle in Paris – to the north side of the house, where he hoped to be laid to rest. 'The beautiful house', remarked the novelist Charlotte Mary Yonge in that same year, 'was like a church in spirit.'

Even as the entire guano supply of Peru was exhausted, Tyntesfield continued to grow, or at least appreciate. In 2002, more than a hundred years after the end of the Guano Age, the estate and its contents were acquired by the National Trust for £24 million. I thought of my own connection with Runcorn's tanneries, which went into decline in the 1950s and 1960s. Even after they spilled out their workers decades ago, my own history was tied to these factories, the hides of cattle, the bark of oak used for tanning and the labour of the men who dipped the hides. Along with tannery scholarships and a cathedral in Medak, India, the profits from Runcorn's tanneries contributed to my education in a provincial English boarding school. If Tyntesfield reeked of the Chinchas, then perhaps it is partly true that this book smells of skin and flesh.

As I walked on Chincha Norte now, it was hard to link this place with Tyntesfield. Unlike the country house, the islands receive few visitors, apart from biologists studying their bird life. Devoid of rainfall, they have no vegetation, no lush gardens where one can sit in the shade; Flaviano had only one plant, a *maracuyá*, or passion fruit, which he tended with almost obsessive care. Rather than the variegated tones of leaded windows, there was the monochrome of guano, as white as marble. For Tyntesfield's chapel, there was one rusted cross. The only parallel I found were the island's bird-stamped windows. On some islands, the guardians had ingeniously converted a translucent plastic sack into a windowpane. On its front was an image of a guanay cormorant, the principal guano-producing bird of Peru. Unlike the stylised images of boobies in Tyntesfield, it accurately reflected the form of the bird, its long neck, dark body and white breast suggesting a hand that was familiar with the creature it depicted. Like a stained-glass window, the sack allowed some light into the hut, bathing the soggy wood in a yellowish glow, the outline of the bird inked on its surface.

Travelling through Peru, from the Andean puna to its lowland coastal areas, I'd often find sacks like these, but not yet emptied, on sale for a mere fifty soles, about fifteen dollars. Normally accompanied by a large sign of a cormorant, they held cheap organic fertiliser from the islands, a counterpoint to the synthetic fertilizer that now dominates the market. When I caught sight of the image of the guanay, I'd instinctively recoil from the sacks, associating them with the Chinchas, the slaughterhouse of men. Yet there were signs that suggested a break with the past, a rupture. I put my hand inside one of the sacks and scooped up some of the powder, then brought my hand upwards so that I could smell it, yet it was not necessary: the smell came out to meet me. I burst out sneezing and retching and sealed the bag, but the genie had long escaped.

During the Guano Age, the guano mined and sold consisted of ancient deposits, whose consistency led some observers to believe that it was a mineral, the 'accumulation of bodies of animals and plants'. Yet there was no mistaking this powder, no possibility of comparing it to a seam of coal. Fishy and putrid, it was clearly fresh from the bird. Stamped across these sacks in bold red capitals was a word, *artesanal*, its meaning somehow jarring against the nature of the contents. I'd come across a similar word many times in Sardinia to describe highly skilled weavers and their patient labour, harvesting fibres and working them into complex patterns. What was the art of guano? Could one be a maestro of excrement?

In Lima, Carla Cepeda, a Peruvian biologist, explained the art of guano extraction to me. Every year she and a team of officials from Agro Rural, a Peruvian state body, select one of the guano islands to be the site of the harvest. 'We only choose places that have not been harvested for a minimum of five years. To avoid the disturbance of the birds, so they can build their nests and the guano can accumulate.' Once an island is chosen, some four hundred men head out to harvest the guano, living in concrete blocks or under tarpaulins. Rising in the early hours to beat the sun, they stock up on a breakfast of oats before digging and bagging the excrement. 'They need to be very tough to do the job,' Carla told me. 'Many have been doing it for generations. I have seen grandparents, parents, and sons.'

In the nineteenth century there was little consideration given to the islands' birds: they were hunted, kept as pets or simply driven away from their homes. In this harvest, unlike those dark times, great attention is paid to the life on the islands. 'If the birds are breeding, then it's impossible to enter the islands. If we find any chicks or eggs, then we stop the harvest and have to find another space.' It can be tiring, frustrating work. The year I visited Peru, Carla had chosen one island, Macabí, as the harvest site, before quickly withdrawing. 'The dilemma was with our friends the penguins, because one day they practically invaded the island and started mating.'

In 1907 a young ornithologist named Robert Coker journeyed to the guano islands. Shortly after completing his PhD at Johns Hopkins, he headed there to study Peru's fisheries at the request of its government. Over a period of a year, he travelled out to the islands by steamer or cargo sloop, observed the birds and camped out in a tent on the guano. 'Unless one is enslaved to the freshwater bath and other "comforts of civilization,"' he wrote, 'a camping experience on one of the Peruvian islands is never to be regretted.' Like a duck farmer, he'd wade into the midst of their wild colonies, enclosing himself among the birds. 'If one awaits motionless and with much patience, the birds, after a while, will return to the nests and gradually close in around the observer.'

At the time of Coker's visit, the islands still bore the scars of the Guano Age: little or no ancient deposits of guano remained and the open coffins of Chinese labourers littered the islands, their bones scattered on the guano. Yet Coker did not dwell for long on the iniquities of the labourer trade, the archaeology of guano extraction or the global market for fertiliser. His gaze was drawn to the islands' living birds: huge colonies of pelicans and guanay cormorants, the colossal living acreage that coats the islands, and its capacity to produce fresh guano. On the island of Chincha Sud, he estimated, there were 180,000 pairs of guanay cormorants, producing 6,400 long tons of excrement per year. 'The fowl which dwelt on the South Chincha Island alone . . . ', he commented, 'might well be regarded as an asset representing a value of several millions of dollars.'

As Coker pointed out, the Incas understood that living birds produced ample amounts of guano. According to Garcilaso de la Vega, the guano-producing birds were protected by the Incas; those who killed them were severely punished. But during the Guano Age there was little consideration of the capacity of living birds, whom Coker called the 'innocent agents in the production of the mines of wealth'. Recounting this history, a strain of anger and frustration runs through his writing. He excoriated earlier travellers, such as Humboldt, who ignored or underestimated the capacity of living birds to produce guano. 'Had Humboldt given correct information,' he lamented, 'it might earlier have been recognised that it was commercially useful to protect the birds as was done in the time before the conquest.'

If Coker's words hint at anger, then they also betray great optimism, namely that the repetition of past mistakes could be avoided through the knowledge of the value of the birds' excrement. Citing the birds' value, he proposed to the Peruvian government that the guano islands be nationalised, a rotational harvest established, and the birds protected. 'It is necessary', he wrote, 'to cease treating the birds as wild animals whose homes men may invade almost like beasts of prey to seize the useful product, regardless of the producing birds. Under a wiser policy the birds will be looked upon as domestic animals, engaged in a useful labour, and from which a greater benefit will be derived the more an intelligent consideration is shown for their welfare.'

After reading so much about the excesses of the Guano Age, I felt drawn to Coker's plans, their boldness and idealism. A mere thirty years old, Coker had been tasked with writing a report on fisheries, but had ended up proposing a new industry, merging conservation and profit. His vision, his references to Incan conservation, must have seemed absurdly idealistic, almost utopian. But much to its credit, the Peruvian government took his advice and nationalised most of the guano-producing islands. A joint stock company was set up – the Compañía Administradora del Guano (CAG) – that gained the sole right to harvest and sell guano. But it was also required that the birds be protected to maximise the guano harvest. Guards, like Flaviano, were placed on the islands to oversee the welfare of the birds, and harvesting was organised on a rotational basis, giving the birds time to recover from the disruption. And gradually the birds returned.

More than one hundred years since Coker's visit, the harvest of fresh guano still takes place on twenty-two islands and eleven headlands. The year that I visited Peru, the island of Mazorca had been chosen as the site of the guano harvest by Agro Rural – the modern-day incarnation of the CAG. Rising precipitously some eighty metres from the ocean, Mazorca is more dramatic than any of the Chinchas, almost resembling the granite peak of an Alpine mountain. Although its guano is produced by the same birds, it smells, looks and feels quite different from that of the Chinchas. Moistened by sea mist and occasional rainfall, Mazorca guano is closer to the sticky brown soil of Nuar's cave than to the dry ammoniacal powder of these three islands that lie farther south.

As I approached Mazorca from a skiff, I saw the authors of the fresh guano, huge black and white stains of guanay cormorants and boobies, on the upper slopes of the island. They coated its steep cliffs and slopes, feathering almost the entire area. We often associate such concentrations of birds with factory farms, but here the cormorants had elected to nest in one huge dense cluster, forming a vast factory of organic fertiliser. Every day the birds tumbled off the cliff to gorge on fish, returned to their nests and dumped their ammonia-rich excrement, baked in the constant equatorial heat. The process never stopped, and if a month could become a moment, the island would have risen as bread, casting a lengthening shadow across the water.

Away from the birds I spotted ghostly figures coated in the yellow powder, hacking into a seam of guano with spades, brushing it into sacks, which they piled into neat pyramids. As they proceeded up the cliff face, the brown soil of the guano was replaced by the bright white of guano-dusted granite. It was as if they were cleaning the island, tending this chunk of rock where nothing would ever grow. A series of images by the renowned Brazilian photographer Sebastião Salgado show the Serra Pelada mine in Brazil, 270 miles south of the mouth of the Amazon. In 1979 a local girl found six grams of gold in a local river, leading to one of the largest gold rushes in the country's history. Thousands descended on the mine in the hope of getting rich, turning it into an anthill. 'No one is taken there by force,' recalled Salgado, 'yet once they arrive, all become slaves of the dream of gold and the need to stay alive. Once inside, it becomes impossible to leave.' Like those miners in Brazil, the guano diggers were covered in dust and mud, bearing heavy loads, but here the comparison ends. Each of them was paid a wage, regardless of the value of the guano they dug up. There was no hope any of the diggers would make it rich, finding a gold nugget weighing nearly 6.8 kilos, as in Serra Pelada. More construction site than gold mine, the island offered stable wages rather than the chance of a lucky strike.

It had rained a few days earlier and the guano was moist and sticky, pungent with the smell of fresh fish. Concerned that the guano might be unstable, the administrators had laid huge quantities of feathers, a waste product from harvested excrement, along the paths, forming a powdered lattice. As the Agro Rural administrators and I walked along the path towards the top of the island, the guano stuck to our boots in great clumps, weighing on them like clods of mud. It was hard work making our way up the steep side of the island, our boots plunging ankle deep into the fresh guano, as unstable as volcanic sand. As we moved upwards, guanay cormorants spilled off the side of the cliff, almost rolling downwards to head off to fish. One of the administrators received a direct hit from a cormorant. '*Plata*,' he said, meaning 'silver', a reference to guano's onetime value. In the city, to be shat on is to have good luck, a consolation for the inconvenience. But with such great concentrations of birds, no one in the guano islands spoke of good fortune; to receive a direct hit was an inevitability.

At the top of the cliff was a small automated lighthouse, covered in velvety algae. Without knowing why, I stroked its surface, irresistibly attracted to it. It was, I realised, the only green on the entire island, a reminder of the mainland, surrounded by this brown soil. We breathed in the air, hoping to catch the sea breeze, but all we got was the smell of ammonia and putrid fish, released by the moisture of the past few days. I thought back to my time in Nuar's cave, where he lived surrounded by birds. Like the men here, he survived in the company of birds, tolerating their guano and the insects. Yet this island seemed far harsher than his cave, its entire landscape almost designed to degrade the human body. 'There is no moss, or lichen, or grass, or twig, or weed, available, or within a hundred miles and more,' explained Alexander Duffield in the nineteenth century. 'Even the sea does not yield a leaf.'

Lacking vegetable patches or domestic animals, Agro Rural administrators must transport tonnes of food from the mainland to Mazorca, where it is prepared by the island's cook, Marleni Ordóñez, who also often happens to be the only woman who lives on the islands for the whole season. At a quiet moment, I sneaked off to see her, walking down the steps to the island's kitchen. In her fifties, she was busy stirring a huge metal vat of potatoes, her head covered with a plastic hairnet, an admirable precaution on an island covered in excrement and ticks. The diggers 'eat a lot', she told me, taking a short break from the vat. 'They are like construction workers.' I'd imagined that it might be intimidating for one woman to live on these islands surrounded by so many men, but Marleni did not look remotely intimidated. 'I'm used to it,' she said, laughing. 'I see them wander around in their underwear the whole time. I'm like a sister or a mother to them.'

In line with Coker's suggestion, the living quarters of the men lay far away from the nesting birds. Crammed into a small section at the northern tip of the island, they slept in temporary shelters made of frames and blue tarpaulins, stained white and brown by the guano. While the birds were free to wander all over the island, the movements of the workers were restricted; confined to a tight space, they defecated in small latrines suspended over the cliff, their own excrement plunging into the water while that of the birds was conserved, treasured. Although I could not talk with them, I knew that theirs was a hard lot, their own rights limited by the needs of the birds. 'We experience a high rate of desertions,' explained one Agro Rural official in 2016, 'despite the fact that for the past three years we have given them high-quality working conditions. They have cable television and photovoltaic light on more than half of the islands. But we cannot affect the birds with antennas or generators. Or with machinery, for that matter. Because the most important thing [with the birds] is care for the environment. No birds, no guano.'

I thought of the Icelandic harvesters, the men and women who gathered down. At the beginning of the twentieth century, Robert

Coker had sought to design a new contract between harvesters and birds, replacing the logic of Tyntesfield with that of eiderdown. Wading through the guano here, I appreciated how hard it is to re-create this balance, this accord with the birds, which have such little tolerance for human presence, a far cry from the 'barn door' birds of the Westfjords. Nor did the harvesters, waist deep in guano, seem to care much for the seabirds, the unfamiliar forms of guanays or boobies. Wage labourers, rather than farmers, the diggers received $430 a month, a fixed salary instead of the profit from sale. In the early years it was common for harvesters to ride roughshod over the birds, upsetting their breeding patterns, an excess now restrained by monitors from SERNANP, a Peruvian conservation agency. The whole system is predicated on control, regulation, rather than a natural balance.

However fragile the accord between these birds and the harvesters, it is unlikely to be replaced anytime soon by the ruthless extraction of the nineteenth century. From 1909 onwards, the German scientists Fritz Haber and Carl Bosch developed a process in which ammonia could be synthesised from the atmosphere, by means of electricity and a metal catalyst. Much like oil refineries, fertiliser plants using the Haber-Bosch process are now part of our own landscape. Take a trip to a garden store to look for fertiliser, and you are unlikely to find any guano at all; the vast majority of fruit, vegetables and seeds that we eat have been fertilised by nitrates conjured from the air rather than from the gullets of birds. As a fertiliser, guano is now a global irrelevance, its meaning largely political and symbolic.

The remarkable discovery of these two chemists solved the pressing problem of how to feed our rapidly growing population, but it also unleashed a series of unintended consequences, from the development of explosives to the widespread eutrophication of waterways, caused by algal blooms. Synthetic fertiliser production also requires a huge amount of energy, derived mostly from the combustion of fossil fuels, particularly natural gas; in 2004 it was estimated that fertiliser production was responsible for up to 1.2 per cent of all anthropogenic greenhouse gas emissions.

It has been suggested that guano might one day provide a solution to these problems. In the early twentieth century a German carpenter named Adolf Winter built a small wooden platform in Walvis Bay, off Namibia's coast, hoping to attract seabirds and harvest their guano. He expanded it over the years, so that by 1938 it covered an area of three football pitches, yielding a steady supply of guano. Although Peru's coastal waters are much deeper than those of Walvis Bay, in theory its guano harvest could be boosted through the construction of floating bird platforms, but, even so, I doubt that it would resolve the pressures of growing populations. As Gregory T. Cushman points out, by 2000 the entire quantity of nitrogen extracted by Peru's guano and nitrate miners in the nineteenth century could be synthesised by Haber-Bosch plants in a mere ten days.

As with tagua, I could not imagine guano turning back the clock, replacing the synthetic alternative that had relegated it to a niche position. Yet I did not feel the sense of disappointment that I experienced when I learned of tagua's limitations, the absurdity of any dream that it might one day substitute plastics. It was enough that the birds of Peru's guano islands could coexist with humans on the same island. However messy and compromised the guano extraction of today, however difficult the lives of the harvesters, the Peruvian government had somehow found a way for birds, men and one woman to live together in a confined space. This small island, a utopia constructed from the ruins of the bacchanalia of the Guano Age, provided some sense of hope.

When I looked out to the ocean, the sheer number of birds spoke of this continued relationship. I saw interminable lines of birds – cormorants, pelicans, and boobies – which spliced the bright blue sky up into sections. In the air these seabirds had an elegance that they sorely lacked on land, as if transformed by the act of flight. Patrolling in long files, the pelicans skimmed the water with their enormous wings. As I watched them track the motion of the liquid below them, I could grasp why the Spanish word for 'pelican', *alcatraz*, may be derived from the Arabic *al-ġattās*, meaning a species of sea eagle. It seemed an unnecessary display of skill to fly so low, akin to feats in an airshow, yet the pelicans derive great benefit from skimming the water, conserving energy because of slight changes in airflow patterns at such low altitude.

High above the pelicans were huge ribbons of guanay cormorants heading out to fish. Seemingly chaotic by comparison with the pelicans', their lines continually broke up and re-formed, sometimes absorbing the missile-like form of a booby, perhaps believing itself a cormorant. Cormorants are so diffuse throughout the world that they often fail to inspire wonder, but looking up at their lines, I admired the whiteness of their breasts, akin to those of Antarctic birds. The intensity of this colour intrigued the ornithologist Robert Cushman Murphy, who spent years studying Peru's seabirds in the 1920s. Normally, he observed, it is cormorants from the far south that have white breasts; the guanay's whiteness at a latitude close to the equator surely indicated the presence of the Humboldt Current, whose cold temperatures extend all the way down South America.

Since ornithologists and travellers first visited Peru's coast in the sixteenth century, they have struggled to convey to others the huge quantities of birds there, often resorting to the language of liquids, fabrics or the weather. They speak of endless ribbons of birds, pools that stain the islands black like oil, or great clouds blocking out the sun for hours on end. So large were the numbers of birds on these islands that they could tire even the most impassioned ornithologist or visitor, prompting disgust or fatigue. 'I have seen them in double millions at a time, swarming in the sky, like insects on a leaf, or vermin in a Spanish bed,' wrote Duffield. 'Now I no longer care for birds.' After several hours of watching, I too grew tired, overwhelmed by the sheer quantities of bird life.

Several months would pass before the island would be cleared, its brown soil replaced by the glinting white of guano-dusted granite. Several months before all the guano would be stuffed into bags and sent hurtling down via cables to waiting barges destined for Callao. After the work was done, the diggers would return to the Andean highlands, the Agro Rural officials to Lima, still smelling of guano. Only the island guardians would remain, keeping an eye on the birds and their excrement.

The work of conservation of the birds, to maximise the guano harvest, was once highly proactive on these islands. Lizards were regularly released to eat the ticks, believed to cause health problems in the birds; birds of prey were hunted in large numbers; the islands were even cleaned to allow more nesting sites. But today the guardians' main responsibility is to protect the guano and birds from thieves and hunters. A more monastic experience is hard to imagine: working in pairs, they rise at six, tour the island with a blank map, filling it with marks from a black pen to indicate the nesting spots of birds, clean their old soggy huts, take the water temperature, fish and prepare meals. 'We island guards always work in silence,' Ricardo Moreno, the oldest guard, told me.

Now that I write this, I think back to Flaviano, the island guardian on Chincha Norte. Some years ago he had left his job as a guano digger to become a guardian, converting his earnings into his gold teeth. He was three days older than me, but he looked many more, his skin leathered by the sun, its rays reflected by the guano. Unused to conversation, he spoke softly, sometimes inaudibly, and found no awkwardness in silences. Like a monk, he followed his routine of cleaning and cooking to the point of punctiliousness. More than efficiency or professionalism, it seemed to betray a certain anxiety, the knowledge that a single displaced beat might throw off his daily rhythm.

In Dino Buzzati's novel *The Tartar Steppe*, a young army lieutenant is sent to a remote outpost overlooking a vast desert, beyond which is said to lie a hostile force of Tartars. As the writer Tim Parks notes, the novel's protagonist, Giovanni Drogo, at first seems to be a medieval knight as he rides off to this distant fort. But when Drogo arrives, he finds the guards spend most of their time locked in repetitive tasks, scanning the desert day and night for signs of an impending invasion. Constantly talked about, but never seen, the existence of the enemy is inferred through a series of omen-like incidents: the appearance of a riderless horse at the walls of the fort, or a mark in the distance of the desert, supposedly a railway supply line.

Buzzati wrote his novel while he was working on night shifts at *Corriere della Sera*, the Italian daily, and felt that time was slipping away from him. 'It often occurred to me', he said, 'that that routine would never end and so would eat up my whole life quite pointlessly.' In his novel, Drogo's life is slowly consumed in a monotonous routine of guarding and inspections. Although he never intends to stay long at the fort, expressing disdain for its repetitive rhythm, he is unable to escape its spell. Time, painfully and inexorably, passes, and Drogo enters into middle and then old age, reflecting on his lost years with longing. When he returns home on a break, he finds that his friends have moved on and hardly noted his absence.

Reading Buzzati's novel, I'd often think about the island guardians. The Spanish word *tiempo* – derived from the Latin *tempus* – means both 'weather' and 'time', one the measure of the other. We often measure time by changes in seasons: April showers, the falling of leaves. But here on the guano islands the weather was constant, unchanging: the birds came and went, the days remained the same length, and it hardly ever rained. If memory works by finding milestones, then the islands offered a uniform surface, unmarked by any variation. In a surreal counterpoint to the temporal chrysalis of Tyntesfield, all the days on the island could fold into one another; an event that happened four days ago and one that happened four months ago seemed equally distant.

The only real change for the island guardians came every three weeks or so when they returned home to visit family. It was a welcome break from the rhythm of the island, but these returns to the mainland could be testing, painful reminders of the parallel lives of their families and friends. While time on the islands seemed to stand still, solid like the guano, it flowed freely on the mainland: governments fell, the guardians' family members got jobs, married and betrayed each other. Flaviano said he returned from the islands one day to find his wife had taken up with someone else. He'd set out for the islands a married man and he returned single, transformed; now he had no one to go back to.

As on the Tartar Steppe, it seemed as if the guardians on the guano islands were constantly waiting for something to happen, some event as great as the Spanish occupation in 1864. One could easily turn the chatter of birds into human voices, perhaps those of guano thieves or the ghosts of long-dead Chinese diggers. But the reality was that there was little risk of any assault, incursion, or theft of guano. There is no ogre, no giant to slay; the greatest challenge was to stay sane in the land of the birds, maintain discipline and tolerate the stillness, the absence of human voices.

If there is a threat to the birds, there is little the guardians can do about it. Since the 1950s, Peru's fishing industry has grown exponentially to become one of the country's most profitable. Countless purse seiners patrol the waters around the guano islands, harvesting the birds' food, the *anchoveta*, most of which is processed into fishmeal to feed livestock. Faced with the seiners, the birds cannot compete, their beaks and feet no match for purse nets. During my trips to the islands I'd watch this conflict play out between the birds and the vessels. At times, great hailstorms, thousands and thousands of boobies, would dive down into the water to fish *anchoveta*, turning the ocean into a bubbling froth. But I knew their haul was a fraction of that of the seiners in the distance, silently mining the waters, their activity only partly visible, like the body of an iceberg. 'Nature does not allow animals to overeat their prey, but no such margin of safety governs the taking of anchoveta for fish meal,' wrote Robert Cushman Murphy as early as the 1950s.

On top of the strains of competing with purse seiners, Peru's seabirds must contend with El Niño – Southern Oscillation (ENSO), a cyclical disturbance that disrupts the Humboldt Current, the upwelling system that supports marine organisms off the coast of South America. Guanay cormorants have proved particularly vulnerable to fluctuations in their food supply and will abandon their nests in droves during El Niño cycles. Even if there is a fall in demand for fishmeal, reducing the activities of seiners, these birds will continue to face great pressure; it has been predicted that El Niño events will increase in frequency as the earth warms.

In the evening on Chincha Norte, Flaviano cooked me a dinner of potatoes and noodles, grateful for company. He spoke little, conserving conversation as if it were the island's freshwater. Afterwards, as the sun set, he suggested we watch television in his room, which looked out over the cliff. His favorite programme, he said, was *La voz Perú*, the Peruvian equivalent of *The Voice*. The format of the show, where hapless contestants compete to secure the attention of judges from the music industry, was familiar, well-worn. Yet as we sat on his bed staring at the old television, the setting could not have been more odd: the human voices were drowned out by a non-human racket: the chattering of birds, the roaring of sea lions and the braying of Humboldt penguins as loud as asses in some demented farmyard. Before long, Flaviano fell asleep, lulled by the contestants' singing, snoring gently.

Sitting with him, his face lit up by the crackling screen, I got a glimpse of the terrible loneliness of these island guardians. Robert Coker wrote that it was necessary to treat the guano birds as domesticated, but the truth was that these men had been wilded living among the birds. While the guano provided for their families, it deprived them of all familial contact, demanding a different husbandry. In the place of their wives, all the guards had were thousands of birds, tonnes upon tonnes of guano, radio transmissions of human voices, and hundreds of pin-ups of female forms, representations of absence. Worried about the solar panel's charge, I moved to turn the television off, but stopped myself at the last moment, leaving it blaring as I returned to my own room.

EPILOGUE

After I left my job at Canary Wharf, giant tunnel-boring machines cut through London's clay and chalk, skirting underground power lines, tunnels, sewers and abandoned tube stations.Working twenty-four hours a day, seven days a week, they gouged twenty-six miles of tunnel to make way for the new Crossrail line. As I gathered objects, archaeologists unearthed and classified artefacts that lay in the borers' pathway: prehistoric flints, a Tudor bowling ball, Roman horseshoes, medieval ice skates made from animal bone, nineteenth-century ginger and jam jars, and a fragment of a mammoth's jawbone, which was found in Canary Wharf itself, a short distance from my desk.

During the Pleistocene, mammoths roamed the British Isles together with other ancient mammals. At that time, the Isle of Dogs did not belong to the London Docklands Development Corporation, but to the mammoth, cave bear and hyena. Unlike their modern-day cousins, these mammals were covered in thick fibre, which protected them from the extremes of the Ice Age. Today we must rely on sophisticated scientific techniques to re-create their appearance, their gait and their calls, but it was not always so. Surviving cave paintings suggest that our ancestors had intimate knowledge of ancient mammals, a familiarity that today characterises only our relationship with domesticated animals.

The discovery of the jaw fragment was greeted with great

excitement, but I found something unsettling in it, a reminder of the fierce debate as to why mammoths and other megafauna no longer roam the earth's surface. Unlike the extinction of the dinosaurs, there is little evidence that the disappearance of these creatures coincided with an asteroid strike or a mass volcanic eruption. No one has conclusively established why they vanished in Europe and elsewhere, but to some scientists it is not a coincidence that many of them met their demise shortly after the arrival of modern humans. 'Though it might be nice to imagine there once was a time when man lived in harmony with nature, it's not clear that he ever really did,' writes Elizabeth Kolbert.

When I got back home from my travels, I placed my objects in a small cabinet. I would take them out, move them around, and compare them, just as Ole Worm might have done with his own collection. When I picked them up, I would imagine their passage, tracing them backwards through many hands until I reached the harvest, the language of finance replaced by that of fox killers, nest collectors, scat scoopers, byssus weavers, fibre shearers, nut foragers and excrement diggers, their voices reverberating inside my head. As I listened to them, I wanted them to tell a different story from that of the mammoth's jaw fragment; I wanted them to serve as proof that our proliferation on the planet did not require the cutting of different branches of the tree of life.

In medieval times, nature or objects from the wild were often treated as mysteries to be decoded. By reading the natural world as a book, one could discern lessons or parables, and come closer to an understanding of God's will. But admiring my own objects, I did not find any commandments, prophecies or imperatives. Unlike the sea silk of the Byssus Museum, they could not be arranged to reveal sacred injunctions. To read them as a map was to ask too much, and yet perhaps they could act as compasses, a means of orientation.

Gavin Maxwell wrote of eiderdown farmers as having 'green fingers', tending their ducks and the land. For all the distortions in his view of the harvest, for all that he romanticised eiderdown, perhaps he touched on the truth. Gardens, writes Michael Pollan, teach us the 'lesson that nature and culture can be compromised, that there might be some middle ground between the lawn and the forest.' At their best, the harvesters taught us a similar lesson, giving and taking, cropping and pruning, identifying their own interest with the world around them.

At moments, I'd found this middle ground in unexpected places, a cave or an island of excrement, but I also appreciated its fragility, its contingency. No longer is the harvest a private endeavour, a small garden tended by its owner; now it is one shaped by the desires of distant others and their small choices, from undoing a button to drinking coffee. No longer do commodities travel in a single line; now, to quote Calvino, they form 'spider-webs of intricate relationships seeking a form'. We must learn to step back and view them, tracing their threads like those of dewed gossamer to see our own lives reflected within them. Only then can we decide whether to keep or leave them. With that thought, I returned my objects to their cabinet, catching a glance of myself in the glass as I did so.

ACKNOWLEDGEMENTS

I'm indebted to my agent, Patrick Walsh, who was the first to imagine how a book might emerge from a short essay on the plumage of a sea duck. It is largely down to his judgement, relentless enthusiasm and encouragement that the essay proved a fragment in a mosaic of stories. After helping me draft my proposal, he found two first-class editors, Paul Slovak at Viking and Stuart Williams at The Bodley Head. Most authors are lucky if they find one great editor: I was fortunate enough to get two. Together Paul and Stuart helped me turn a collection of object biographies into the layered text that I wanted to write. At a time when publishers face large pressures (of which writers often have no inkling), they treated me with great kindness, tolerating delays and questionable drafts. As a first-time author, I will always be grateful to them.

I am fortunate to have friends and family who thought that this undertaking was worthwhile. Although you won't find their names in the endnotes, three friends in particular shaped this book, often over long walks in Skye, Jura, the Norfolk Broads and the Brecon Beacons. After I returned from Iceland, Matt Lloyd-Rose pushed me to write up my notes on the eiderdown trade and commented, prophetically, that it might make a book one day. Since then he has edited countless drafts and acted as a judicious sounding board. Henry Eliot read several early drafts, infusing the project with his contagious enthusiasm for literature and collecting. At low points I found it helpful to recall his absolute commitment to the value of the *eccentric* (literally that which lies beyond the centre). Andy Wimbush, himself an experienced critic, tore apart my proposal; his dedication to clarity of expression sets a standard to which I aspire.

Thank you to my family, stepfamily and in-laws, who helped me out in so many ways – a roof over my head in London and Somerset, research material on family history, company during cold-water swims in Dover and an unlimited supply of flapjacks – and to Mel Akhurst, whose quiet support over more than three decades served me well during this project. Bill Higgins, Tom Hodgson, Tobie Mathew, John McManus and Rick Sowerby, who are all familiar with the challenges of writing, helped keep my feet on the ground as I struggled to put pen to paper. In London and elsewhere I'm grateful for the friendship and support of Charles Arrowsmith, Gill Barnes, Martina Caldana, Jack Castle, Alessandra Chessa, Alessandro Chiozzi, Piero Cornice, Una Dimitrijevic, Ferdinand Eibl, Katharina Franz, Alexander Goldsmith, Ally Goldsmith, Georgie Gould, Alice Hamlett, Evelyn Heathcoat Amory, Mary Henes, Lydia Lloyd-Rose, Chris Moses, Cecily Motley, Laura Pitel, Nick Plested, Sarah Ramsey, Philip Rosenbaum, Brett Scott, Belinda Sherlock and Ollie Watts.

Thank you to my former colleagues in London whose red pens tightened my writing; the labour of report writing served as a good apprenticeship for this book. In addition to sponsoring the prize that kickstarted this project, the *Financial Times* has offered encouragement and editorial advice; I'm grateful to Isabel Berwick and Alice Fishburn, two judicious editors of my early journalistic forays. Jason Gathorne-Hardy kindly invited me to be the writer-in-residence at the Alde Valley Spring Festival, a welcome boost at a time when I had done very little writing. Stephanie Cronin, Carrie Donald, Victoria Glendinning, Eugene Rogan and Xan Smiley gave helpful advice on publishing and research at a time when I was floundering. I am grateful to them all.

Midway through writing this book I moved to Philadelphia, a city that welcomed me and my wife with warmth and openness. Thank you to all our new friends, who have taught us so much about life in the United States; to the staff of the Christian Street YMCA, which provided welcome respite from the solitude of writing; and to the Penn Museum, especially Adria Katz, keeper of its Oceanian Section, who showed me Furness's collection of objects. At Princeton University, I owe a particular debt to the staff of the Firestone and Marquand libraries, where much of this book was written. A special thank you to Irena Grudzińska Gross, who welcomed us so warmly in Princeton and in New York.

While researching this book I spoke to hundreds of people who were willing to share details of their lives, trades and expertise. My thanks to the harvesters, the men and women whose stories form the backbone of this book, and who perhaps had the least to gain from talking to me. I'm also indebted to the specialists who took the time to educate me on an array of subjects, from echolocation to domestication. Special thanks to the readers who looked over specific chapters, often with short notice: Henrik Balslev, Grischa Brokamp, Alessandra Chessa, Lord Cranbrook, Lesley Kinsley, Kjell Larsson, Gabriela Lichtenstein, Felicitas Maeder, Andri Snær Magnason, Jonathan Morris, Robert O'Hagan, Gigliola Sulis, Julie Velásquez Runk and Jane Wheeler. If any errors have crept in, I am solely responsible; I ask their forgiveness and that of the reader.

EIDERDOWN. In Iceland, I'm indebted to the harvesters: Fjölnir Ás-björnsson; Salvar Baldursson and family; Björn Baldursson, Ingunn Sturlu-dóttir and Baldur Björnsson; Erla Friðriksdóttir; Katrín Sigríður Alexíusdóttir, Alexíus Jónasson, Magnús Jónasson and family; Guðrún Gauksdóttir; Valdi-mar Gíslason and Edda Arnholtz; Andri Snær Magnason; Björgvin Sveinsson; Jón Sveinsson and Ólína Jónsdóttir; and Zófonías Þorvaldsson. Special thanks to Jón, who proved an inexhaustible source of knowledge on eiderdown, and to Andri, whose own writing on natural resources sparked many ideas in this book and provided inspiration. I'm also grateful to Petr Glazov, Alexandra Go-ryashko, Alexander Kondratyev, Helmut Kruckenberg, Kjell Larsson and Ivan Mizin, all of whom shared their expertise on waterfowl and/or the Arctic. I owe a particular debt to Alexandra, who has done so much to clarify what we know about the history of down harvesting.

EDIBLE BIRDS' NESTS: I'm grateful to Lord Cranbrook for opening up the world of nest harvesting. Without his contacts, advice and zest, it would not have been possible to scratch the surface of this opaque trade. In Sarawak, I owe a debt to Hasbullah Abdulrahman, Haidar Ali and the Forest Depart-ment Sarawak, Lucy Bulan, Nuar bin Haji Jaya and Lim Chan Koon. My thanks to my expedition companions in Sabah and Java: Sarah Ball, Chiwon Chin, Jamie Curtis Haywood and Alan Speck; and to Aida Rahman and Mohd Salleh bin Sabti for assistance on Mantanani. In Java, I'm grateful to Anton Hoo, Boedi Mranata, Ariani Mranata, Harry Mranata and Johannes Siegfried. In Johor Bahru, my thanks to Tan Boon Siong and Siah Ching Hoon for showing me around the city's bird houses and explaining the process of nest cleaning. While in Kuala Lumpur, I benefited from conversations with Harry 'Swiftlet' and Dato Tan Chee Hong. Thanks to Will and Sal Addington, who kindly hosted me in KL at a time when I still reeked of bat guano. Craig Thorburn provided helpful advice; his research on swiftlets, politics and do-mestication paved the way for this chapter.

CIVET COFFEE. I'm grateful to Tony Wild for sharing his deep knowledge of coffee and civets; and to Anna Nekaris, who generously invited me to Cipaganti as a guest of Project Little Fireface (nocturama.org); my thanks to all the project's staff and volunteers, particularly Robert O'Hagan, who looked after me while I was laid low with a terrible bug. Chris Shepherd and Peter Roberts kindly an-swered questions about the civet trade. In Bali, I benefited from discussions with Kerry Negara, whose documentary *Done Bali* still provides an excellent intro-duction to the island more than two decades after it was filmed. My thanks to Anton Muhajir, who introduced me to Agung Alit, Degung Santikarma and Taman 65, all of whom welcomed me warmly in Bali. Matthew Ross kindly shared information on his harvesting techniques, while Jonathan Morris gener-ously answered questions about the history of coffee. I'm indebted to them all.

SEA SILK. At the Byssus Museum and its defence group, I'm grateful to Karl Logge, Giuseppe Mongiello, Fabrizio Steri and Chiara Vigo. I'm indebted to Sant'Antioco's other sea silk weavers, among them Giuseppina and Assuntina Pes, as well as those who have endeavoured to shed light on the material's forgotten history: Ignazio Marrocu and the staff of Sant'Antioco's Ethnographic Museum, Claudio Moica and Antonella Senis. Felicitas Maeder was an invaluable and patient source of information and support, while Rubens D'Oriano, a Sardinian archaeologist, provided helpful background on the intersection of archaeology and fantasy in Sardinia. I also benefited from conversations with Sarah Caronni, Glenda Giampaoli, Arianna Pintus and Emma Diana, who welcomed me into her home and shared much information about her father. Finally, thanks to all my friends in Sardinia, who have always welcomed me with such warmth to their island: Massimo Aresu, Giuseppe Littera, Pablo Sole and Gigliola Sulis.

VICUÑA FIBRE. My gratitude to Romina Chullen and José Salvador Palacios for hosting me in Lima. In Ayacucho, my thanks to all the vicuñeros who shared their stories: Manuel Cabrera, Héctor Galván, Reino Joyo, José Sarmiento and Hernán Sosaya; and to all the shearers, led by Abraham Yarihuamán Huamaní (Canchito). I'm indebted to Allan Flores Ramos and the SERNANP staff and volunteers for their hospitality at Pampas Galeras. Special thanks to the community of Lucanas, particularly Norma Bujaico, José Héctor Quispe Mitma, Corina Rojas Escajadillo, Francisco Yarihuamán Huamaní and Marco Zúñiga. I benefited from conversations and/or correspondence with a range of experts and businesspeople: Adolfo Bottari, Amy Cox Hall, Barbara Fraser, John Hemming, Nils Jacobsen, Gabriela Lichtenstein, Saxon Logan, Gustavo Lozada, Alfonso Martínez, Wilfredo Pérez Ruiz, Catherine Sahley, Omar Siguas, Beatriz Torres and Jane Wheeler. My thanks to them all, particularly Jane, who spent so much time explaining camelid evolution, genetics and morphology; and Alfonso, who spoke so frankly about the trade in fibre.

TAGUA. My thanks to Catherine Ankerson and Lucio Arevalo, who kindly hosted me in Quito and provided an oasis of support and stability during my research. In Manabí, I'm indebted to Franklin Pilay and his family for welcoming me to La Crucita and showing me their tagua palms. In Manta and Jipijapa, I was assisted by Maritza Cárdenas, Mario Cedeño, Fredy Heredia, Pedro Loaiza, Alan Posliga and Isaac Saenz, all of whom have detailed knowledge of tagua cutting and trading. In Manta, Klaus Calderón, Fernanda Zanchi and Francisco Luna kindly shared their knowledge on the history of tagua, while Tatiana Hidrovo Quiñónez and José Elías Sánchez Ramos provided helpful background on the history of the city. Henrik Balslev, Grischa Brokamp and Julie Velásquez Runk kindly shared their botanical expertise.

GUANO. I'm indebted to the island guardians: Jesús Pérez (Islas Ballestas) and Flaviano Huanri Cochachin and Jorge Tarazona Paredes (Chincha Norte), as well as Manuel Díaz and Ricardo Moreno (now retired). During the guano harvest on Mazorca, I was assisted by Ivan Balbín, Melchor Llica, Jack Robinson, Victor Ropón and Leonela Valdivia. I could not have visited the islands without the permission and assistance of Peruvian state officials: my gratitude to Tomás Cedamanos, Carla Cepeda and Cinthia Irigoin at Agro Rural; and Mariano Valverde, Luis Cortez and Jorge Vásquez López at SERNANP. Thanks to the fishermen, Richard and Fito, who took me out to the islands and treated me no differently from how they treat the guardians they normally ferry; to Susana Cárdenas-Alayza and Marco Cardeña, who showed me around the Punta San Juan Reserve; and to the following experts on bird life and/or guano: Irma Franke, Lesley Kinsley, Daniela Lainez del Pozo and Carlos Zavalaga. Gaelle Fisher assisted with German-language translation, while the Scientific Society Swakopmund and Sue Ogterop retrieved historical documents.

In New York, a big thank you to Andrea Schulz at Viking, who backed this project from the outset. Under Paul Slovak's patient eye, Viking turned my manuscript into a book with consummate professionalism. I'm indebted to Anna Jardine, production editor, who had the good sense to commission Angelina Krahn to carry out a copyedit; Anna and Angelina eliminated a terrifying number of mistakes, ranging from zoological nomenclature to my baffling insistence on repeating the word 'cinnamon'. They are masters of their craft. I'm grateful to Elizabeth Yaffe and Gretchen Achilles, who designed the book's elegant cover and interior, respectively; to Allie Merola, Paul Slovak's assistant, who did much work behind the scenes; and to Carolyn Coleburn, Viking's publicist.

In the UK, I have received a huge amount of support from the lovely people at The Bodley Head and Vintage; thank you to Anna-Sophia Watts, Emmeline Francis, and Lauren Howard for their patient assistance; to Julia Connolly and Lily Richards for conceiving and designing such a sensuous cover; to Aidan O'Neill for publicity; and to Monique Corless for assistance with foreign rights. I owe a particular debt to Stuart Williams, publishing director at The Bodley Head, whose decision to launch an annual essay prize ultimately led me to take this interesting and unexpected path. Finally, my gratitude to everyone at – or formerly at – Conville & Walsh, especially Carrie Plitt, who provided invaluable assistance with my proposal; and to John Ash, Patrick Walsh's able assistant at PEW Literary Ltd.

It's typical to write that a book could not have been finished without one's partner. In the case of this book, it never would have been started, let alone completed, without the support of my wife, Gabriella. Long before I ever expressed any interest in writing about nature, she gently pushed me to go to Iceland to pursue the story of eiderdown, telling me that my interest was not an aberration, but something that spoke to an important part of myself. Since then she has acted as an unpaid researcher, editor and translator, nurturing the book to the extent that it often felt like a joint project. Even though it feels as if your name should be on the cover, it is dedicated to you with all my love.

NOTES

INTRODUCTION

4 **gathered objects from around the world:** For background on the history of collecting, I relied on Paula Findlen, *Possessing Nature: Museums, Collecting, and Scientific Culture in Early Modern Italy* (Berkeley: University of California Press, 1996).

4 **'In the world of the curiosity cabinet':** Mark A. Meadow, 'Hans Jacob Fugger and the Origins of the Wunderkammer', in *Merchants & Marvels: Commerce, Science, and Art in Early Modern Europe*, ed. Pamela H. Smith and Paula Findlen (New York: Routledge, 2002), 185.

5 **My favourite wunderkammer:** The catalogue to Worm's collection, the *Musei Wormiani historia* (Leiden: Ex Officina Elseviriorum, 1655), can be viewed online on the website of the Biodiversity Heritage Library and the Internet Archive.

5 **Worm conserved these things, he wrote:** Ole Worm to Arngrímur Jónsson, June 20, 1639, in H. D. Schepelern and Holger Friis Johansen, eds., *Breve fra og til Ole Worm* (Copenhagen: Munksgaard, 1967), vol. 2, 132, trans. and quoted in Valdimar Tr. Hafstein, 'Bodies of Knowledge: Ole Worm & Collecting in Late Renaissance Scandinavia', *Ethnologia Europea* 33, no. 1 (January 2003): 9. In addition to Hafstein, I relied on H. D. Schepelern, *Museum Wormianum: Dets forudsætninger og tilblivelse* (Copenhagen: Wormianum, 1971); Jole Shackelford, 'Documenting the Factual and the Artifactual: Ole Worm and Public Knowledge', *Endeavour* 23, no. 2 (1999): 65–71.

5 **During the sixteenth and seventeenth centuries:** See Findlen, *Possessing Nature*.

5 **Worm was influenced by these collectors:** Shackelford, 'Documenting the Factual and the Artifactual', 65.

5 **'the clear intention to lead':** Worm, *Musei Wormiani historia*, trans. and quoted in H. D. Schepelern, 'The *Museum Wormianum* Reconstructed: A Note on the Illustration of 1655', *Journal of the History of Collections* 2, no.1 (January 1990): 84.

6 **others spoke of these complex new trade routes:** Hafstein, 'Bodies of Knowledge', 8, 10.

7 **'to establish the relationships':** Italo Calvino, *Invisible Cities*, trans. William Weaver (New York: Harcourt Brace Jovanovich, 1974), 76. I am indebted to Mark Meadow for drawing the parallel between the curiosity cabinet and Calvino's story. See 'Hans Jacob Fugger and the Origins of the Wunderkammer', 185.

EIDERDOWN

13 **Over centuries eiderdown has been treasured:** On the history of eiderdown as a trade commodity, see Birgitta Berglund, '*Fugela Federum* in Archaeological Perspective – Eider

Down as a Trade Commodity in Prehistoric Northern Europe', *Acta Borealia*, 26, no. 2 (2009): 125–28.

13 **The Vikings apparently filled:** Carla J. Dove and Stephen Wickler, 'Identification of Bird Species Used to Make a Viking Age Feather Pillow', *Arctic* 69, no. 1 (March 2016): 29–36.

14 **means of payment:** Berglund, '*Fugela Feðerum,*' 128.

14 **Romanov tsars . . . had a taste:** During her 1888 visit to Russia, Emma Blackstock, an American traveller, commented that the tsar's couches and chairs were 'stuffed with eiderdown'. See Emma Moulton Frazer Blackstock, *The Land of the Viking and the Empire of the Tsar* (New York and London: G. P. Putnam's Sons, 1889), 112. Although travellers may describe 'eiderdown' in Romanov palaces, it is hard to know whether it really was down from an eider; in the nineteenth century, as today, the word *eiderdown* was often used to refer to any filling of goose or duck down. (On the usage of the word, see page 245.)

14 **According to trade bodies:** See, for instance, FAQs, International Down and Feather Bureau (IDFB), accessed November 4, 2018, www.idfb.net/faqs.

14 **reported to be common:** Oliver Milman, '"Ethical Down": Is the Lining of Your Winter Coat Nothing but Fluff?', *The Guardian*, January 14, 2016.

15 **Parish ministers were among:** Ásthildur Elva Bernhardsdóttir, *Learning from Past Experiences: The 1995 Avalanches in Iceland* (Stockholm: Försvarshögskolan, 2001), 45.

15 **mothers may starve to death:** Chris Waltho and John Coulson, *The Common Eider* (London: T & A D Poyser, 2015), 36.

16 **'so strong,' recorded one Belgian eider enthusiast:** Johan Beetz, 'Notes on the Eider', trans. Charles W. Townsend, *The Auk* 33, no. 3 (July 1916): 287. The English nature writer Gavin Maxwell was less disdainful of the liquid's odour: 'It is a curious smell, very pungent, and resembling the smell of frying liver. For the few who may have smelled the cooking liver of a stag after the rut has begun the simile is almost exact. It is a warm, perhaps hot odour, suggesting its own colour of rich brown; most humans do not find it unpleasant, but feel that if it were increased to the least degree it would be nauseating.' See Gavin Maxwell, *Raven Seek Thy Brother*, in *Ring of Bright Water: A Trilogy* (Boston: Nonpareil Books, 2011), 272.

16 **'The earthen walls that surrounded it':** C. W. Shepherd, *The North-West Peninsula of Iceland: Being a Journal of a Tour in Iceland in the Spring and Summer of 1862* (London: Longmans, Green, 1867), 104, quoted in S. F. Baird, T. M. Brewer and R. Ridgway, *The Water Birds of North America* (Boston: Little, Brown, 1884), vol. 2, 74, quoted in Charles W. Townsend, 'A Plea for the Conservation of the Eider', *The Auk* 31, no. 1 (January 1914): 17.

17 **'The cooing notes, so long few':** Townsend, 'A Plea for the Conservation of the Eider', 20.

18 **untouchable, a sacred bird:** Legislation to protect eiders dates back to the centuries after Iceland's settlement. According to Robin Doughty, 'incipient protection' is found in the Jónsbók (c. 1281), one of the foundations of Icelandic legislation; complete legal protection was later introduced by the Danes in 1847. See Doughty, 'Eider Husbandry in the North Atlantic: Trends and Prospects', *Polar Record* 19, no. 122 (May 1979): 447–59.

18 **The Norwegians once harvested:** Although eiderdown harvesting is not widespread in Norway, the practice still takes place on the Vega Archipelago, a group of islands off the coast. On the history of harvesting in Norway, see Berglund, '*Fugela Feðerum*'; Bente Sundsvold, '*Stedets herligheter* – Amenities of Place: Eider Down Harvesting through Changing Times', *Acta Borealia* 27, no. 1 (2010): 91–115.

18 **It is considering:** According to Landsvirkjun, the National Power Company of Iceland, proposals to lay a submarine cable date back sixty years. In 2016 a joint UK – Iceland task force published a feasibility study on the cable, but at the time of writing neither government has committed to the project; see 'Overview of IceLink', on the Landsvirkjun website, accessed January 6, 2019, www.landsvirkjun.com/researchdevelopment/research/submarine cabletoeurope.

18 **The future of Iceland's renewable energy sources:** For a critical take on Iceland's energy policy, see Andri Snær Magnason, *Dreamland: A Self-Help Manual for a Frightened Nation* (London: Citizen Press, 2012).

20 **'fairy tale told backward':** Rebecca Solnit, 'News from Nowhere: Iceland's Polite Dystopia', *Harper's Magazine*, October 5, 2008.

20 **Icelandic word for 'windfall' is *hvalreki*:** Magnason, *Dreamland*, 67.

20 **'Abundant fishing grounds have been a resource':** Email from Jón Sveinsson, July 14, 2014. I have lightly edited his text for clarity.

21 **It is said that Saint Cuthbert:** For a more detailed look at the saint's relationship with eiders, see Mark Cocker and Richard Mabey, *Birds Britannica* (London: Chatto & Windus, 2005), 103.

21 **embellishment or pure fantasy:** According to Antone Minard, Saint Cuthbert's association with eiders began almost five centuries after his death 'in a piece of twelfth-century *folklorismus*'. See Minard, 'The Mystery of St Cuthbert's Ducks: An Adventure in Hagiography', *Folklore* 127, no. 3 (December 2016): 325–43.

21 **'barn door' bird:** Richard F. Burton, *Ultima Thule: Or, A Summer in Iceland* (London: William P. Nimmo, 1875), vol. 2, 46.

21 **'tame as horse-pond geese':** Ibid., 112.

21 **'No salute must be fired':** Ibid., 45.

24 **travelled to Iceland on a research trip:** Richard Frere, *Maxwell's Ghost: An Epilogue to Gavin Maxwell's Camusfearna* (London: Victor Gollancz, 1976), 99, 101; Douglas Botting, *Gavin Maxwell: A Life* (London: Eland, 2017), 449–50.

24 **he also adored eiders, writing that:** Maxwell, *Raven Seek Thy Brother*, 270.

25 **he dreamed of re-creating the success:** Maxwell's memoir touches only briefly on his eider project, but his private papers set out his plans in great detail. See File of Correspondence Relating to a Proposed Eider Duck Colony Project, 1965–69, Papers of Gavin Maxwell, National Library of Scotland.

25 **'Maxwell – Queequeg with an Oerlikon':** Robert Macfarlane, 'Shark Attack: Gavin Maxwell's Harpoon at a Venture', *The Guardian*, July 19, 2014.

25 **'an elaboration of fluttering flags':** Maxwell, *Raven Seek Thy Brother*, 272.

26 **'seem to have "green fingers"':** Gavin Maxwell, 'Eider Duck Colonies', undated memo, Papers of Gavin Maxwell.

26 **'Iceland on the brain':** Burton, *Ultima Thule*, vol. 1, ix–x.

27 **It is now believed:** Páll Hersteinsson et al., 'Elstu þekktu leifar melrakka á Íslandi', *Náttúrufræðingurinn* 76 (2007), 13–21 (English summary, 20).

27 **a myth emerged that they had been sent:** See William Hooker, *Journal of a Tour in Iceland in the Summer of 1809* (Yarmouth, UK: J. Keymer, 1811), 42–43.

27 **dates back to 1295:** Páll Hersteinsson et al., 'The Arctic Fox in Fennoscandia and Iceland: Management Problems', *Biological Conservation* 49, no. 1 (1989): 67–81.

27 **Famed throughout the Westfjords:** Valdimar and his fellow fox hunters are the subject of a short documentary by the Icelandic filmmaker Haukur Sigurðsson, *Skolliales* (2012).

29 **more than four thousand ravens per year:** Kristinn H. Skarphéðinsson et al., 'Breeding Biology, Movements, and Persecution of Ravens in Iceland', *Acta Naturalia Islandica* 33 (1990): 1. Skarphéðinsson recently warned that the killing of ravens has led to their serious decline; see Jóhann Bjarni Kolbeinsson, '3.000 hrafnar veiddir á hverju ári á Íslandi', RÚV, September 6, 2017, accessed January 6, 2019, http://www.ruv.is/frett/3000-hrafnar-veiddir-a-hverju-ari-a-islandi.

29 **disappearance of Iceland's white-tailed eagle:** Staff, 'The Icelandic Sea Eagle Population Larger Than Any Time Since 19th Century', *Iceland Magazine*, October 23, 2017, accessed January 6, 2019, https://icelandmag.is/article/icelandic-sea-eagle-population-larger-any-time-19th-century.

29 **the 'dark side' of the eiderdown trade:** Email from Andri Snær Magnason, November 18, 2018.

30 **When settlers first came:** For background on the settlement period, I relied on Jesse Byock, *Viking Age Iceland* (London: Penguin, 2001).

31 **much of early Icelandic law:** See, for example, Vicki Szabo's study *Monstrous Fishes and the Mead-Dark Sea: Whaling in the Medieval North Atlantic* (Leiden: Brill, 2008), 243–75.

31 **'The fighters kept exchanging':** *Grettir's Saga*, trans. Denton Fox and Hermann Pálsson (Toronto: University of Toronto Press, 1974), 20.

32 **'serious doubts . . . whether you can simply transfer':** Harry Milne to Gavin Maxwell, November 2, 1965, Papers of Gavin Maxwell.

32 **'I am going ahead':** Gavin Maxwell to George Waterston, October 7, 1965, Papers of Gavin Maxwell.

32 **permission from the National Trust for Scotland:** Maxwell, *Raven Seek Thy Brother*, 273.

32 **budget of £2,100:** Undated budget, Papers of Gavin Maxwell.

32 **Two of his assistants:** Frere, *Maxwell's Ghost*, 230.

32 **'to be fascinated by the sounds':** Ibid., 231.

32 **'a virtual reproduction of an Icelandic colony':** Ibid., 230.

32 **the thousands of ducks:** Although Frere refers to the breeding programme as a failure, Maxwell did apparently increase his eider population in 1968. According to Douglas Botting, who visited him in October of that year, Maxwell claimed that there were two hundred nests on the island, a notable increase from his original estimate of thirty breeding pairs. See Botting, *Gavin Maxwell*, 530.

32 **'We hoped that spring would bring':** Ibid., 230.

33 **'The magic that had once glossed':** Maxwell, *Raven Seek Thy Brother*, 290.

34 **'The islands of Vigr and Œdey':** Shepherd, *The North-West Peninsula of Iceland*, 102, quoted in Baird, Brewer and Ridgway, *The Water Birds of North America*, vol. 2, 74, quoted in Townsend, 'Conservation of the Eider', 17.

35 **jackets filled solely with Æðey's eiderdown:** Staff, '66 North Makes 8,000 Dollar Eiderdown Jackets to Order', *Iceland Magazine*, July 14, 2016, accessed January 6, 2019, https://iceland-mag.is/article/66-north-makes-8000-dollar-eiderdown-jackets-order.

36 **'In that rough climate':** Georges-Louis Leclerc, *Histoire naturelle des oiseaux* (Paris: Imprimerie Royale, 1783), vol. 9, 109, quoted in translation in James Rennie, *The Architecture of Birds* (London: Charles Knight, 1831), 75.

38 **'All these plants':** Michael Pollan, *The Botany of Desire: A Plant's-Eye View of the World* (New York: Random House Trade Paperback, 2002), xv.

39 **'From the time of the ravages of the Black Death':** Richard F. Tomasson, *Iceland: The First New Society* (Minneapolis: University of Minnesota Press, 1980), 18.

39 **another country with abundant numbers:** The history of eiderdown in Russia is hard to reconstruct because of the paucity and fragmentary nature of sources. I am grateful to Dr Alexandra Goryashko, a Russian eider specialist, for sharing her knowledge of harvesting in Russia, a subject on which there is precious little English-language research. On the history of Russia and eiderdown, see, for example, her articles 'Obyknovennaya istoriya gagi obyknovennoi' [A common history of the common eider], *Khimiya i zhizn* [Chemistry and life] 5 (2002): 32–35; 'Gagachii pukh, znakomy i neznakomy' [Eiderdown, familiar and unfamiliar], GoArctic.ru, accessed January 6, 2019, https://goarctic.ru/live/gagachiy-pukh-znakomyy-i-neznakomyy.

39 **'The unlucky creatures, ceaselessly harassed':** Franz Ul'rikh, *Kemskii uezd i rybnye promysly na Murmanskom beregu* [Kemskii Province and the fishing industry on the coastline of Murmansk] (Saint Petersburg: V. Kirshbaum Press, 1877), quoted in Goryashko, 'Gagachii pukh'.

40 **During Soviet times:** On eiderdown harvesting in Soviet Russia, see Alexandra Goryashko, 'Attempts to Establish Eider Farms in the USSR, and Why These Failed', *Environment &*

Society, Arcadia 14 (Summer 2017), accessed January 6, 2019, http://www.environmentand
society.org/arcadia/attempts-establish-eider-farms-ussr-and-why-these-failed.

41 '**The Eider is a conservative bird**': Quoted from a draft of Alexandra Goryashko's work on eiderdown farms in the USSR, which she shared with me on November 18, 2016.

41 '**The composition of the harvester**': Email from Jón Sveinsson, February 22, 2018. I have
lightly edited his text for clarity.

EDIBLE BIRDS' NESTS

45 '**weirdest bird's-nesting expedition**': Apsley Cherry-Garrard, *The Worst Journey in the World*
(New York: Penguin Books, 2005), 228.

45 '**looked about thirty years older**': Sara Wheeler, *Cherry: A Life of Apsley Cherry-Garrard* (London: Jonathan Cape, 2001), 118.

46 '**We are a nation of shopkeepers**': Cherry-Garrard, *The Worst Journey in the World*, 564.

46 '**golden fleece, or the skin**': Robert Fraser, *Victorian Quest Romance: Stevenson, Haggard,
Kipling and Conan Doyle* (Plymouth, UK: Northcote House, 1998), 5.

46 '**probably the most primitive bird in existence**': Cherry-Garrard, *The Worst Journey in the
World*, 227–28. I am indebted to Elizabeth Leane's analysis of *The Worst Journey in the World*
and the quest narrative: 'Eggs, Emperors and Empire: Apsley Cherry-Garrard's "Worst Journey"
as Imperial Quest Romance', *Kunapipi*, 31, no. 2 (January 2009): 18–34.

47 **Like forensic investigators**: J. R. Green, 'The Edible Bird's-Nest, or Nest of the Java Swift
(Collocalia nidifica)', *Journal of Physiology* 6, no. 1–2 (April 1865): 40–45.

47 '**to be composed of fine filaments**': Sir George Staunton, *An Authentic Account of an Embassy
from the King of Great Britain to the Emperor of China* (London: W. Bulmer, 1797), vol. 1, 287,
quoted in Rennie, *Architecture of Birds*, 293.

47 **Others speculated that they were built**: Ibid., 288–306.

47 **travelled to the state of Sarawak**: To recount the Furness-Hiller expedition, I have relied on
William Henry Furness, *The Home-Life of Borneo Head-Hunters: Its Festivals and Folk-Lore*
(Philadelphia: J. B. Lippincott, 1902); Adria H. Katz, 'Borneo to Philadelphia: The Furness-
Hiller-Harrison Collections', *Expedition* 30, no. 1 (1988): 65–72; and Matthew Schauer, 'A
Beautiful Savage Picture: Adventure Travel, Ethnology, and Imperialism in Nineteenth-
Century Borneo', in *Exploring Travel and Tourism: Essays on Journeys and Destinations*, ed.
Jennifer Erica Sweda (Newcastle upon Tyne: Cambridge Scholars, 2012), 7–28.

48 '**They intend making a journey**': Unnamed Honolulu newspaper, c. October 28, 1895, quoted
in Katz, 'Borneo to Philadelphia', 66.

48 **it was ruled by Charles Brooke**: On the history of Sarawak, see Steven Runciman, *The White
Rajahs: A History of Sarawak from 1841 to 1946* (Cambridge, UK: Cambridge University Press,
1960).

48 '**there are various tribes indigenous to the soil**': Charles Hose, *Fifty Years of Romance and
Research; Or, A Jungle-Wallah at Large* (London: Hutchinson, 1927), 39.

48 **Despite the efforts of the Brookes**: Katz, 'Borneo to Philadelphia', 67; Furness, *The Home-Life
of Borneo Head-Hunters*, 69–70.

49 '**missing link**' **between humans and apes**: Earl of Cranbrook, 'The "Everett Collection from
Borneo Caves" in the Natural History Museum, London: Its Origin, Composition and Potential for Research', *Journal of the Malaysian Branch of the Royal Asiatic Society* 86, no. 304 (June
2013): 79–112.

49 '**Insensate, indeed**': Furness, *The Home-Life of Borneo Head-Hunters*, 181.

50 '**It is the ideal forest primeval**': Ibid., 173.

50 '**Gentle, simple-hearted creatures**': Ibid., 174.

50 '**It seemed the veritable entrance of the Inferno**': Ibid., 181.

50 'The extent beyond, in the utter darkness': Ibid., 182.

51 'Our presence and the echoing of our voices': Ibid., 182.

51 *Aerodramus maximus*: In relating the ecology and biology of swiftlets, I rely on Lim Chan Koon and Earl of Cranbrook, *Swiftlets of Borneo: Builders of Edible Nests* (Kota Kinabalu, Malaysia: Natural History Publications [Borneo], 2002).

51 'perceived the roof of the cavern': Pierre Poivre to Georges-Louis Leclerc, in Georges-Louis Leclerc, *The Natural History of Birds* (London: A. Strahan et al., 1793), vol. 6, 572, quoted in Rennie, *The Architecture of Birds*, 291.

52 'There is a membranous tube': Sir Everard Home, 'Some Account of the Nests of the Java Swallow, and of the Glands that Secrete the Mucus of Which They Are Composed', *Philosophical Transactions of the Royal Society of London* (London: W. Bulmer, 1817), 335, quoted in Rennie, *The Architecture of Birds*, 297.

52 'Up these poles': Furness, *The Home-Life of Borneo Head-Hunters*, 182.

53 when the Chinese first started: On the history of human use of the nests, I relied on Lim and Cranbrook, *Swiftlets of Borneo*, 61–63; Bien Chiang, 'Market Price, Labor Input, and Relation of Production in Sarawak's Edible Birds' Nest Trade', in *Chinese Circulation: Capital, Commodities, and Networks in Southeast Asia*, ed. Eric Tagliacozzo and Wen-Chin Chang (Durham, NC: Duke University Press), 407–31; Leonard Blussé, 'In Praise of Commodities: An Essay on the Cross-Cultural Trade in Edible Bird's-Nests,' in *Emporia, Commodities, and Entrepreneurs in Asian Maritime Trade c.1400–1750*, ed. Roderich Ptak and Dietmar Rothermund (Stuttgart: Franz Steiner Verlag, 1991), 317–35.

53 the first actual reference to the nests: According to Lim and Cranbrook, this can be found in the publication *Yin-shih hsü-chih* (What We Need to Know About Food and Drink). See *Swiftlets of Borneo*, 62.

53 Zheng He brought back birds' nests: Ibid., 62.

53 became a major export commodity: Chiang, 'Market Price, Labor Input, and Relation of Production', 411.

53 the nests are or were believed: Amy Lau and David Melville, *International Trade in Swiftlet Nests* (Cambridge, UK: TRAFFIC International, 2004), 5.

53 'birds' nests were thought to be': Blussé, 'In Praise of Commodities', 323.

54 'Anything *less* like good food': Tom Harrisson, 'Birds and Men in Borneo', in Bertram Smythies, *The Birds of Borneo* (Edinburgh: Oliver and Boyd, 1960), 32.

54 one food scientist: Massimo Marcone, 'Characterization of the Edible Bird's Nest "Caviar of the East"', *Food Research International* 38, no. 10 (December 2005): 1125–34.

54 'I am . . . of the opinion': Francesco Redi, *Experimenta circa res diversas naturales, speciatim illas, quae ex Indiis adfernatur* (Amsterdam: Sumptibus Andreae Frisii, 1675), 166, quoted in translation in Rennie, *The Architecture of Birds*, 289–90.

54 'Properly cooked the edible bird's nest': Albert M. Reese, 'Unusual Human Foods', *Scientific Monthly* 14, no. 5 (May 1922): 477.

55 'the most primitive type of human being': William Furness, 'Observations on the Mentality of Chimpanzees and Orangutans', *Proceedings of the American Philosophical Society* 55, no. 3 (1916): 281.

55 spent years trying to teach orangutans: Furness records his experiments in his presentation to the American Philosophical Society; see ibid., 281–90.

55 trickle of swiftlet saliva: Chinese communities in West Coast cities in the United States and Canada account for 2 per cent of the nest market, according to an email from Lord Cranbrook to me, November 20, 2018.

57 Sarawak's forest was logged: For background on land use in Niah, see Tina Hansen, 'Spatio-Temporal Aspects of Land Use and Land Cover Changes in the Niah Catchment, Sarawak, Malaysia', *Singapore Journal of Tropical Geography* 26, no. 2 (July 2005): 170–90.

58 'tree crash, cicada buzz': Tom Harrisson, *World Within: A Borneo Story* (London: Cresset Press, 1959), 3.

59 'damp caves inhabited by snakes': G. H. R. von Koenigswald, *Meeting Prehistoric Man* (London: Thames & Hudson, 1956), 78, quoted in Tom Harrisson, 'The Great Cave of Niah: A Preliminary Report on Bornean Prehistory', *Man* 57 (November 1957): 166.

59 'If there ever were early cave men': Harrisson, 'The Great Cave of Niah', 161.

59 a 'larder,' offering 'relief from want': Barbara Harrisson, 'Tom Harrisson's Unpublished Legacy on Niah', *Journal of the Malaysian Branch of the Royal Asiatic Society* 50, no. 1 (1977): 44.

61 'Trees and fruits': Alfred Russel Wallace, *The Malay Archipelago: The Land of the Orang-Utan and the Bird of Paradise* (London: Macmillan, 1922), 58.

61 'Every now and then someone falls': Harrisson, 'Birds and Men in Borneo', 31.

61 'I've gone to great extremes': 'Malaysia', series 2, episode 3, of *Gordon's Great Escape*, directed by Tom Coveney et al., featuring Gordon Ramsay, aired on Channel 4, May 23, 2011.

62 set out to document the harvest of birds' nests at Niah: See Judith M. Heimann, *The Most Offending Soul Alive: Tom Harrisson and His Remarkable Life* (Honolulu: University of Hawai'i Press, 1998), 314; 'Birds' Nest Soup', episode 4 of *The Borneo Story*, directed by Hugh Gibb, narrated by Tom Harrisson, aired on BBC Television, October 20, 1957.

62 'ran the operation': Sir David Attenborough's words are taken from the excellent documentary *Tom Harrisson: The Barefoot Anthropologist*, produced by Icon Films, presented by Sir David Attenborough, aired on BBC 4, January 18, 2007.

62 'living among strange people': *Who's Who 1975* (London: A. and C. Black, 1975), quoted in Heimann, *The Most Offending Soul Alive*, 1. Details on Harrisson's biography also come from Heimann.

63 Harrisson's first major book: *Savage Civilization* (New York: Alfred A. Knopf, 1937). In his introduction, Harrisson writes: 'I had been taught that cannibals were terrible, and completely inferior to ourselves. Yet I reached the position of identifying myself, in sentiment, with the cannibals rather than with the whites who have tried to "civilise" them.'

66 It depicts a human figure: Gene Kritsky, *The Quest for the Perfect Hive: A History of Innovation in Bee Culture* (Oxford: Oxford University Press, 2010), 11–12.

66 'honey was so precious': Bee Wilson, *The Hive: The Story of the Honeybee and Us* (London: John Murray, 2004), 5.

66 an accord existed: On the history of cave management at Niah, see Lim and Cranbrook, *Swiftlets of Borneo*, 106.

66 the harvest of nests is strictly controlled: I am grateful to Lim Chan Koon for sharing his unpublished management plan. 'Mini Co-Management Model: A Proposal to Manage Small to Moderate Size Black-Nest Swiftlet Caves Scattered at Subis and Sekaloh Areas', undated.

67 'There is probably no other part': Harrisson, 'Birds and Men in Borneo', 21.

68 start of the birds' nest harvest: See ibid., 31–32. Benedict Sandin recounts alternative origin stories in 'Some Niah Folklore and Origins', *Sarawak Museum Journal* 8, no. 12 (1958): 646–62, summarised in Chiang, 'Market Price, Labor Input, and Relation of Production', 417–18.

68 70 per cent of the harvest in Sarawak: Earl of Cranbrook, 'Report on the Birds' Nest Industry in the Baram District and at Niah, Sarawak', *Sarawak Museum Journal* 33, no. 54 (January 1984): 145–70, cited in Quentin Gausset, 'Chronicle of a Foreseeable Tragedy: Birds' Nests Management in the Niah Caves (Sarawak)', *Human Ecology* 32, no. 4 (August 2004): 488.

69 a million pairs of swiftlets: Harrisson, 'Birds and Men in Borneo,' 30.

69 started to dig into its guano: See Heimann, *The Most Offending Soul Alive*, 321–25.

69 proved to be roughly correct: On the deep skull at Niah, see Chris Hunt and Graeme Barker, 'Missing Links, Cultural Modernity and the Dead: Anatomically Modern Humans in the Great Cave of Niah (Sarawak, Borneo)', in *Southern Asia, Australia, and the Search for Human*

Origins, ed. Robin Dennell and Martin Porr (Cambridge, UK: Cambridge University Press, 2014), 90–107.

71 **'become thick with calling birds'**: Lord Medway (now Lord Cranbrook), 'Cave Swiftlets', in Smythies, *The Birds of Borneo*, 63.

71 **the consumption of birds' nests**: David Jordan, 'Globalisation and Bird's Nest Soup', *International Development Planning Review*, 26, no. 1 (2004): 98. Jordan does not provide a source for this statement, though the image of Mao dining on birds' nests does seem as likely as Lenin nestling under an eiderdown duvet.

72 **'It does not matter whether the cat'**: Deng, quoted in Bin Zhao, 'Consumerism, Confucianism, Communism: Making Sense of China Today', *New Left Review*, no. 222 (March – April 1997): 43–44.

72 **often compared to that of silver:** See, for example, John Crawfurd, *History of the Indian Archipelago: Containing an Account of the Manners, Arts, Languages, Religions, Institutions, and Commerce of Its Inhabitants* (Edinburgh: Archibald Constable, 1820), vol. 3, 434, quoted in Rennie, *The Architecture of Birds*, 304–5.

72 **the retail price of a kilo of cleaned black nests:** Lim and Cranbrook, *Swiftlets of Borneo*, 97. The precise ranges they cite for the cost of a kilo of black nests are RM (Malaysian ringgit) 5,300 to RM 7,100 ($1,378 to $1,846), and RM 23,500 to RM 26,000 ($6,110 to $6,670) for a kilo of white nests. I have calculated dollar values using the conversion rate for January 1, 1999, found at XE.com.

72 **'are a more useful barometer':** Lim and Cranbrook, *Swiftlets of Borneo*, 97.

72 **In the depths of the Great Cave:** On the destruction at Niah, see Gausset, 'Chronicle of a Foreseeable Tragedy', 487–507; Joseph Hobbs, 'Problems in the Harvest of Edible Birds' Nests in Sarawak and Sabah, Malaysian Borneo', *Biodiversity & Conservation* 13, no. 12 (November 2004): 2209–26.

74 **'It'd be aimed':** Email from Andri Snær Magnason, August 9, 2016.

75 **'by nature limited and incapable of being augmented':** Crawfurd, *History of the Indian Archipelago*, 437, quoted in Rennie, *The Architecture of Birds*, 306.

76 **References to beekeeping . . . ancient Egypt:** Kritsky, *The Quest for the Perfect Hive*, 11–13.

76 **believed to be the oldest examples ever discovered:** Amihai Mazar and Nava Panitz-Cohen, 'It Is the Land of Honey: Beekeeping at Tel Reḥov', *Near Eastern Archaeology* 70, no. 4 (December 2007): 202–19.

76 **the origins of the white nests:** On the history of swiftlet farming, see Lim and Cranbrook, *Swiftlets of Borneo*, 142–49; Craig Thorburn, 'The Edible Birds' Nest Boom in Indonesia and South-East Asia', *Food, Culture & Society* 17, no. 4 (2014): 535–53; Craig Thorburn, 'The Edible Nest Swiftlet Industry in Southeast Asia: Capitalism Meets Commensalism', *Human Ecology* 3, no. 1 (2015): 179–84.

76 **In London, great campaigns:** Colin Barras, 'Why Cities Are Unleashing Birds of Prey into Their Skies', BBC Earth, October 4, 2016, accessed January 6, 2019, http://www.bbc.com/earth/story/20161003-why-cities-are-unleashing-birds-of-prey-into-their-skies; 'Feeding Trafalgar's Pigeons Illegal', BBC News, November 17, 2003, accessed January 6, 2019, http://news.bbc.co.uk/2/hi/uk_news/england/london/3275233.stm; Hugh Muir, 'Hawks Do Their Worst but Cost of Pigeon War Is Problem for Mayor', *The Guardian*, September 29, 2006.

77 **The life of the zoologist Lord Cranbrook:** For a summary of Cranbrook's career, see Geoffrey W. H. Davison, Hoi Sen Yong and David R. Wells, 'Cranbrook at Eighty: His Contributions So Far. Ornithologist, Mammalogist, Zooarchaeologist, Chartered Biologist and Naturalist', *Raffles Bulletin of Zoology*, supplement, *Cranbrook at Eighty: His Contributions So Far*, 29 (2013): 1–7.

80 **Dr Boedi as the King of Birds' Nests:** 'Our Founder', Our Stories, Pristine Farms Premium Bird Nest, accessed January 6, 2019, http://pristine-farms.com/en/our-stories.

83 **they can order a batch of queens online:** See, for example, 'VSH-Italian Queen Bees', Wild-flower Meadows, accessed January 6, 2019, https://wildflowermeadows.com/queen-bees-for-sale/.

83 **swiftlets must be seduced:** For a flavour of techniques used to draw birds, see the website of the swiftlet farming consultant 'Harry Swiftlet', accessed January 6, 2019, http://swiftletfarming.blogspot.com.

83 **a great proliferation of theories:** Wilson, *The Hive*, 238.

84 **the result of nitrifying bacteria:** See Thorburn, 'The Edible Birds' Nest Boom', 538, and 'The Edible Nest Swiftlet Industry', 182.

86 **ever-growing body of research . . . possible health benefits of birds' nests:** Thorburn, 'The Edible Birds' Nest Boom,' 536, and 'The Edible Nest Swiftlet Industry,' 179.

87 **'The chief peculiarity in my hives':** Lorenzo Langstroth, *A Practical Treatise on the Hive and the Honey-Bee* (New York: C. M. Saxton, 1857), 15. See Wilson, *The Hive*, 224–25; and Kritsky, *The Quest for the Perfect Hive*, 110–19.

87 **a result of collective malaise:** On the ecological threats facing swiftlet populations, see Lim and Cranbrook, *Swiftlets of Borneo*, 135–36; and Thorburn, 'The Edible Birds' Nest Boom', 543–44, and 'The Edible Nest Swiftlet Industry', 181.

89 **'I have seen a kestrel flying':** George Orwell, 'Some Thoughts on the Common Toad', in *Some Thoughts on the Common Toad* (London: Penguin Books, 2010), 3.

90 **all taurine cattle may be descended:** Ruth Bollongino et al., 'Modern Taurine Cattle Descended from Small Number of Near-Eastern Founders', *Molecular Biology and Evolution* 29, no. 9 (September 2012): 2101–4.

90 **Przewalski's horse . . . the feral descendant:** Charleen Gaunitz et al., 'Ancient Genomes Revisit the Ancestry of Domestic and Przewalski's Horses', *Science*, February 22, 2018.

91 **in house farms . . . little genetic variation:** On the genetics of swiftlets, see Thorburn, 'The Edible Birds' Nest Boom', 546; and Earl of Cranbrook et al., 'The Species of White-Nest Swiftlets (Apodidae, Collocaliini) of Malaysia and the Origins of House-Farm Birds: Morphometric and Genetic Evidence', *Forktail* 29 (2013): 78–90.

91 **One of the putative ancestors . . . grey-rumped swiftlet:** Earl of Cranbrook, 'Genomics of Edible-Nest Swiftlets: A 20th Century Domestication', unpublished proposal for a multinational research project.

91 **check on their bird populations:** Saul Mallinson et al., 'A Brief Account of the Present State of Sabah's Birds'-Nest Caves and the Conservation Status of Edible-Nest Swiftlets in Sabah', *Sabah Society Journal* 32 (2015), 77.

96 **19 per cent of the export revenue:** Ibid., 74.

96 **the wild salmon, whose numbers continue to plummet:** I am indebted to Craig Thorburn for his consideration of the parallels between swiftlet farming and aquaculture. See Thorburn, 'The Edible Birds' Nest Boom', 546–47.

CIVET COFFEE

101 **In 1692, Daniel Defoe faced financial ruin:** Theodore Newton, 'The Civet-Cats of Newington Green: New Light on Defoe,' *Review of English Studies* 13, no. 49 (January 1937): 10–19; Maximillian E. Novak, *Daniel Defoe: Master of Fictions: His Life and Ideas* (New York: Oxford University Press, 2001), 93–96; and Robert Kuttner, *Debtors' Prison: The Politics of Austerity Versus Possibility* (New York: Alfred A. Knopf, 2013), 173–207.

101 **'It is greater then any Cat':** Edward Topsell, *The Historie of Foure-Footed Beastes* (London: William Iaggard, 1607), 756–57; also quoted in Karl H. Dannenfeldt, 'Europe Discovers Civet Cats and Civet', *Journal of the History of Biology* 18, no. 3 (Fall 1985): 414. Topsell in turn was quoting the English physician John Caius writing to the naturalist Conrad Gesner (also known as Konrad Gessner).

102 'was first supposed of the cat [genus]': John Hill, *An History of Animals* (London: Thomas Osborne, 1752), 556, quoted in Dannenfeldt, 'Europe Discovers Civet Cats and Civet', 423.

102 By 1821, John Edward Gray had settled: John Edward Gray, 'On the Natural Arrangement of Vertebrose Animals', *London Medical Repository* 15 (1821): 301.

102 some thirty-eight species: *Mammal Species of the World: A Taxonomic and Geographic Reference*, ed. Don Wilson and DeeAnn Reeder (Baltimore: Johns Hopkins University Press, 2005), vol. 1, 548–58.

102 Its smell, variously compared: Descriptions of civets' smell vary enormously. According to one perfume enthusiast, it has 'notes of honey, leather, wet clay, butter, a fine aged Parmesan or Asiago cheese. . . . It's not disgusting, really, just very, very ripe, earthy and round . . . like the smell of Brie mixed with honey, or like month-old rotten fruit (have you ever fed a bucket of kitchen slop to pigs? . . .)'. See rasputin (Central Texas), January 11, 2013, reply to Sanzio, 'Civet smell', Basenotes, accessed January 6, 2019, http://www.basenotes.net/threads/175128 -Civet-smell.

102 'There has been a time when the produce': William Fordyce Mavor, *Natural History for the Use of Schools* (London: Phillips, 1800), 89, found also, edited, in Christopher Plumb, *The Georgian Menagerie: Exotic Animals in Eighteenth-Century London* (London: I.B. Tauris, 2015), 72. For the history of civet in Georgian England, see Plumb, 72–79.

102 'Give me an ounce of civet': William Shakespeare, *King Lear*, act 4, scene 6, quoted in Dannenfeldt, 'Europe Discovers Civet Cats and Civet', 426.

103 the Sienese writer Pier Andrea Mattioli proposed: Pier Mattioli, *Commentarii in libros sex Pedacii Dioscoridis Anazarbei De medica materia* (Venice: In Officina Erasmiana, apud Vincentium Valgrisium, 1554), 43, cited in Dannenfeldt, 'Europe Discovers Civet Cats and Civet', 424–25.

103 By Georgian times, civet had become: On the history of civet in Georgian England, see Plumb, *The Georgian Menagerie*, 72–79.

103 to yield £2 worth a year: Newton, 'The Civet-Cats of Newington Green', 12.

103 'fitted & made convenient and proper only': Ibid., 13.

104 confined in narrow hutches: On the techniques for harvesting civet, see Plumb, *The Georgian Menagerie*, 74–75.

104 'One [servant] drew the chain': Pietro Castelli, *De hyaena odorifera zibethum gignente quae civetto vulga appelatur* (Messina: Vidua Jo. Franci. Bianco, 1638), quoted in translation in John Jonston, *A Description of the Nature of Four-Footed Beasts* (Amsterdam: Printed for the widow of John Jacobsen Schipper, and Stephen Swart, 1678), 117, quoted in Dannenfeldt, 'Europe Discovers Civet Cats and Civet', 420.

104 'And all your courtly civet cats': Alexander Pope, *The Works of Alexander Pope Esq.* (London, Dodsley, 1738), quoted in Plumb, *The Georgian Menagerie*, 77.

104 'To the credit of taste and elegance': Mavor, *Natural History*, 89, quoted in Plumb, *The Georgian Menagerie*, 77.

105 chemical fixatives now dominate the market: It is difficult to know much civet is used in perfume today. Although Chanel reportedly stopped using civet in 1998, it is still harvested in Ethiopia for export, predominantly to France, where it is used in perfumes. See Emmanuel Do Linh San et al., 'Civettictis civetta [African civet]', *The IUCN Red List of Threatened Species*, 2015; and Patrick House, 'The Scent of a Cat Woman: Is the Secret to Chanel No. 5's Success a Parasite?', *Slate*, July 3, 2012, accessed January 6, 2019, https://slate.com/technology/2012/07 /chanel-no-5-a-brain-parasite-may-be-the-secret-to-the-famous-perfume.html.

105 Cultuurstelsel, or Cultivation System: See Jonathan Morris, *Coffee: A Global History* (London: Reaktion Books, 2019), 86–87.

105 farmers instead used civet scat: See Colin Cahill, 'Feral Natures and Excremental Commodities: Purity, Scale, and the More-than-Human in Indonesia' (PhD diss., University of

California, Irvine, 2017), 25–26. Cahill's work, as far as I am aware, is the only comprehensive study of civet coffee; it provided useful background for this chapter.

105 'selects only the ripest': Thomas Horsfield, *Zoological Researches in Java, and the Neighbouring Islands* (London: Kingsbury, Parbury, & Allen, 1824), n.p., quoted in Cahill, 'Feral Natures and Excremental Commodities', 46.

106 processing of non-civet coffee: See Morris, *Coffee*, 25–27.

106 carries out the same task: Antony Wild, *Coffee: A Dark History* (ebook, Wild Books, 2013); Morris, *Coffee*, 28.

106 Whatever the truth of its origin stories: Cahill, in 'Feral Natures and Excremental Commodities', has scoured the historical record for references to civet or waste beans (45–49). Although he identifies a handful of short references in travel narratives and zoological studies, it is notable that civet coffee does not feature in the major coffee texts of the nineteenth or twentieth centuries, such as William Ukers's 1935 study *All About Coffee* (email from Jonathan Morris, December 14, 2018). As with the trade in birds' nests, it is difficult to ascertain when civet coffee first appeared and whether it was consumed locally or intended for export.

106 'musty' or 'earthy' with 'a top note of rich, dark chocolate': Paul Watson, 'Coffee at a Price Difficult to Digest', *Los Angeles Times*, July 13, 2007.

106 civet coffee is often said to have a low acidity: Morris, *Coffee*, 29.

106 'If it has four legs': Prince Philip is widely reported to have made this comment at a meeting of the World Wildlife Fund (WWF) in 1986. See unnamed author, 'Long Line of Princely Gaffes', BBC News, March 1, 2002, accessed January 6, 2019, http://news.bbc.co.uk/2/hi/uk/1848553.stm.

107 'master champagne maker . . . collecting the very best grapes': Civets are commonly described in such terms, but the truth is more nuanced. See Cahill, 'Feral Natures and Excremental Commodities', 44.

108 The Dutch East India Company initially chose Java: Ibid., 30.

110 eiderdown and edible birds' nests: Worm, *Musei Wormiani historia*, 310–11.

110 a single coffee bean: Ibid., 189.

110 the first actual description of a bean: Schepelern, *Museum Wormianum*, 145; also see Wild, *Coffee: Dark History*, 165.

110 coffee was still a curiosity in Europe: Steven Topik, 'Coffee as a Social Drug', *Cultural Critique* 71 (Winter 2009): 81–106; and Wild, *Coffee: A Dark History*, 143–44.

110 grown commercially on four continents: Morris, *Coffee*, 7.

110 'No longer primarily the beverage': Topik, 'Coffee as a Social Drug', 98.

110 its trade is controlled: It is hard to find up-to-date figures on the trade in green coffee. As of 2013, Fairtrade International, an NGO, reported that three firms control almost half the European coffee trade. See 'Powering Up Smallholder Farmers to Make Food Fair: A Five Point Agenda', Fairtrade International, May 2013.

111 it is robusta . . . that is gaining ground: Wild, *Coffee: A Dark History*, 78–81.

113 fashioned the image of Bali: Adrian Vickers, *Bali: A Paradise Created* (Singapore: Tuttle, 2012).

114 Tony Wild imported a single kilogram: On the recent history of civet coffee, see Wild, *Coffee: A Dark History*, 11–32; and Wild, 'Civet Coffee: Why It's Time to Cut the Crap', *The Guardian*, September 13, 2013. Although Tony Wild claims responsibility for eliciting interest in civet coffee in the 1990s, he was not the only one dabbling in the trade at the time; see Cahill, 'Feral Natures and Excremental Commodities', 44–45.

114 'Tasting *kopi luwak* had become': Wild, *Coffee: A Dark History*, 24–25.

114 Danish merchants began to export eiderdown: It is difficult to know exactly when the trade in Icelandic eiderdown began. Burton writes that English merchants traded eiderdown as early as the fifteenth and sixteenth centuries (*Ultima Thule*, vol. 1, 202), although the earliest reference

to eiderdown in English apparently dates to 1774 (*Oxford English Dictionary Online*, s.v. 'eider-down', accessed January 5, 2019, http://www.oed.com/view/Entry/59968? redirected From=eiderdown&). By the nineteenth century, travel writers make regular references to the trade in Icelandic down.

115 **a strange regression to the practices of Defoe:** In 2016 researchers from World Animal Protection and Oxford University published a paper assessing the living conditions of caged civets in Bali according to guidelines adapted from the World Association of Zoos and Aquariums. After visiting sixteen plantations and observing forty-eight wild-caught palm civets, the researchers concluded, on the basis of eight variables, that the creatures were subject to 'poor welfare'. They fared particularly poorly when it came to diet, the availability of water, and social interactions; see Gemma Carder et al., 'The Animal Welfare Implications of Civet Coffee Tourism in Bali', *Animal Welfare* 25 (May 2016): 199–205. As Wild put it, 'The naturally shy and solitary nocturnal creatures suffer greatly from the stress of being caged in proximity to other luwaks, and the unnatural emphasis on coffee cherries in their diet causes other health problems too; they fight among themselves, gnaw off their own legs, start passing blood in their scats, and frequently die'; see Wild, 'Civet Coffee.'

115 **'I feel as if long ago':** Wild, 'Civet Coffee.'

115 **documentary with the BBC:** 'Coffee's Cruel Secret', episode of *Our World*, presented by Chris Rogers, featuring Tony Wild, produced by Guy Lynn, aired on BBC World News, September 13, 2013.

115 **fluctuations in the price of staple crops:** See also Cahill, 'Feral Natures and Excremental Commodities', 40–41.

117 **'He [Defoe] worked the modern "Statue of Liberty"':** Newton, 'The Civet-Cats of Newington Green', 18.

117 **'While civets might eat':** See Cahill, 'Feral Natures and Excremental Commodities', 44.

SEA SILK

125 **'In one sense, the message':** Alastair Sooke, 'The Man Who Destroyed All His Belongings', BBC Culture, July 14, 2016, accessed January 6, 2019, http://www.bbc.com/culture/story /20160713-michael-landy-the-man-who-destroyed-all-his-belongings.

126 **calling itself enviro-capitalism:** See Terry L. Anderson and Donald R. Leal, *Free Market Environmentalism* (Boulder: Westview Press, 1991); and *Enviro-Capitalists: Doing Good While Doing Well* (Lanham, MD: Rowman & Littlefield, 1997).

126 **I wanted to learn from a material:** While writing this chapter I drew heavily on Project Sea-silk (http://www.muschelseide.ch/en/projekt.html), a research project based at the Natural History Museum of Basel run by the Swiss scholar Felicitas Maeder. She generously shared many primary sources with me, which otherwise I would not have been able to track down. I am also indebted to Helen Scales, a British marine biologist, who in 2015 was the first writer to clearly lay out the nuanced history of sea silk in Sant' Antioco; see her excellent book *Spirals in Time: The Secret Life and Curious Afterlife of Seashells* (London: Bloomsbury Sigma, 2015). I follow in her footsteps.

126 **'The golden-brown sheen':** Conrad Malte-Brun, 'Sur la pinne-marine et sur les tissus fabriqués avec la laine de ce coquillage', *Journal de l'Empire*, 1806, quoted and translated in 'National Fairs and World Exhibitions', Project Sea-silk, accessed January 6, 2019, http://www .muschelseide.ch/en/geschichte/neuzeit/ausstellungen.html.

127 **'burnished gold on the back':** Henry Swinburne, *Travels in the Two Sicilies: In the Years 1777, 1778, 1779, and 1780* (London: T. Cadell and P. Elmsly, 1790), vol. 2, 79.

127 **'fleeces' that were 'obtained from the sea':** Tertullian, *De pallio* (III, 6) quoted and translated in 'Ancient World', Project Sea-silk, accessed January 6, 2019, http://www.muschelseide .ch/en/geschichte/antike.html.

127 *'coquillages à soye'*: René Antoine Ferchault de Réaumur, 'Observations sur le coquillage appellé "pinne marine" ou "nacre de perle," à l'occasion duquel on explique la formation des perles', *Mémoires de l'Académie Royale des Sciences*, Paris (1717), 177 (trans. mine).

127 **'All sea creatures'**: René Antoine Ferchault de Réaumur, 'Des differentes manières dont plusieurs espèces d'animaux de mer s'attachent au sable, aux pierres, et les uns aux autres', *Mémoires de l'Académie Royale des Sciences*, Paris (1711), 108 (trans. mine). For an English summary of Réaumur's investigations, see Clinton G. Gilroy, *The History of Silk, Cotton, Linen, Wool, and Other Fibrous Substances* (New York: C. M. Saxton, 1853), 174–84.

127 **'catch them in the act'**: For a detailed discussion of Réaumur's ingenious methods, see Mary Terrall, *Catching Nature in the Act: Réaumur and the Practice of Natural History in the Eighteenth Century* (Chicago: University of Chicago Press, 2014).

127 **'I saw them feel out their tips'**: Réaumur recounts this phase of his experiment in *Des differentes manières*, 115–18 (trans. mine).

128 **'Spiders and caterpillars' . . . 'the founder who casts metal'**: Réaumur, 'Des differentes manières', 123 (trans. Gilroy, *The History of Silk*, 177).

128 **he might have compared byssus formation to injection moulding**: I owe the comparison to Helen Scales; see *Spirals in Time*, 147.

128 **eliciting hope among chemists**: The biochemist Herbert Waite at UC Santa Barbara has carefully studied the properties of mussel byssus. See, for example, Kathryn Coyne, Xiao-Xia Qin and Herbert Waite, 'Extensible Collagen in Mussel Byssus: A Natural Block Copolymer', *Science* 277, no. 5333 (September 1997): 1830–32.

129 *Pinna nobilis*, **or noble pen shell**: On the biology of the *Pinna nobilis*, see 'Biology', Project Sea-silk, accessed January 6, 2019, http://www.muschelseide.ch/en/biologie.html.

129 **'grow upright out of the bottom'**: Aristotle, *Complete Works of Aristotle: The Revised Oxford Translation*, ed. Jonathan Barnes (Princeton, NJ: Princeton University Press, 1995), vol. 1, 864.

129 **'broadly speaking, the entire genus'**: Aristotle, *Complete Works of Aristotle*, 922.

129 **Giuseppe Saverio Poli studied**: On Poli, see Ilya Tëmkin, 'The Art and Science of *Testacea utriusque Siciliae* by Giuseppe Saverio Poli', in *Atti del Bicentenario: Museo Zoologico 1813–2013*, ed. Maria del Re, Rosanna del Monte and Maria Ghiara (Naples: Centro Musei delle Scienze Naturali e Fisiche, 2015), 147–68.

130 **references to sea silk in ancient to modern texts**: For primary-source citations, see 'Ancient World', Project Sea-silk, accessed January 6, 2019, http://www.muschelseide.ch/en/geschichte /antike.html.

130 **'a pair of curious gloves'**: Autograph Letter Signed (ALS) from Lord Nelson to Emma Hamilton, March 18, 1804, in *The Hamilton and Nelson Papers, 1798–1815*, ed. Alfred Morrison (Printed for Private Circulation, 1894), vol. 2, 226.

131 **'sea-boots, an otterskin cap'**: Jules Verne, *Twenty Thousand Leagues Under the Sea* (New York: Pollard and Moss, 1888), 223.

131 **Attilio Cerruti did experiment with pinna aquaculture**: Lucia D'Ippolito, 'Fra antiche tradizioni e ambizioni industriali: La produzione di bisso marino a Taranto', in *Bisso marino – Fili d'oro dal fondo del mare*, ed. Felicitas Maeder, Ambros Hänggi and Dominik Wunderlin (Milan: 5 Continents Editions, 2004), 73–113. Cerruti published his findings in two articles: 'Primi esperimenti di allevamento della Pinna "Pinna nobilis L." nel Mar Piccolo di Taranto', *La Ricerca Scientifica* 1, no. 16 (1938): 339–47; and 'Ulteriori notizie sull'allevamento della "Pinna nobilis L." nel Mar Piccolo di Taranto,' *La Ricerca Scientifica* 18 (1939): 1110–24.

131 **harvesting . . . cleaning, spinning and weaving**: On the production of sea silk, see 'Production of Sea-Silk', Project Sea-silk, accessed January 6, 2019, http://www.muschelseide.ch/en /verarbeitung.html.

131 **he sent a batch of sea silk to a textile factory**: Giuseppe Basso-Arnoux, *Sulla pesca ed utilizzazione della 'Pinna nobilis' e del relativo bisso* (Rome: Ministero dell'Industria, del Commercio e del Lavoro, 1916), 5.

131 **local practice, confined mainly:** Felicitas Maeder, 'Landscape of Sea-Silk: Traces of Traditional Production Around the Mediterranean Sea', in *Museums and Cultural Landscapes: Proceedings of the ICOM Costume Committee Annual Meeting*, ed. Johannes Pietsch, Milan, July 3–7, 2016.

131 **'What is visibly decreasing':** Harrisson, 'Birds and Men in Borneo', 33.

132 **'The women of that land':** Basso-Arnoux, *Sulla pesca ed utilizzazione*, 6 (trans. mine).

132 **Margherita Sitzia, a sea silk weaver:** Tito Siddi, 'Sfogliando negli archivi della memoria di Sant'Antioco: Margherita Sitzia ed Efisia Murroni, due indimenticabili tessitrici di bisso', *Tottus in Pari*, July 17, 2016.

133 **'I picked away'... 'There was not a shadow':** All quotations are from the obituary of Marjorie Courtenay-Latimer in *The Daily Telegraph*, May 19, 2004. On the coelacanth, see Samantha Weinberg, *A Fish Caught in Time: The Search for the Coelacanth* (New York: HarperCollins, 2000).

134 **When writers and filmmakers:** See, for example, Max Paradiso, 'Chiara Vigo: The Last Woman Who Makes Sea Silk', BBC News Magazine, September 2, 2015, accessed January 6, 2019, https://www.bbc.com/news/magazine-3369178; and Susanna Lavazza, *From Darkness to Light: Marine Byssus and Chiara Vigo* (Bologna: Cartabianca, 2012).

134 **'animals constituted the first circle':** John Berger, *About Looking* (New York: Vintage International, 1991), 3.

136 **Many tales have been told about sea silk:** For a dissection of these stories, see Daniel McKinley, 'Pinna and Her Silken Beard: A Foray into Historical Misappropriations', *Ars Textrina* 29 (June 1998): 9–223.

136 **Chinese traders spoke of aquatic sheep:** Ibid., 67–75.

136 *abu qalamun:* Ibid., 75–94.

136 **byssus to make their sacred vestments:** See Redazione Fame di Sud, 'Chiara Vigo, ultima sacerdotessa della millenaria arte del bisso: Viaggio nell'universo della seta marina', *Fame di Sud*, April 3, 2014, accessed January 6, 2019, http://www.famedisud.it/chiara-vigo-ultima -sacerdotessa-della-millenaria-arte-del-bisso-viaggio-nell'universo-delle-seta-marina/.

136 **'Remember when it says that King Solomon':** Angela Corrias, 'Weaving Sea Silk in Sardinia: Preserving an Ancient Art', *GoNOMAD*, 2010 (precise date unknown), accessed January 6, 2019, https://www.gonomad.com/3046-weaving-the-silk-of-the-sea-in-sardinia.

136 **'Once collected, the sea silk':** Stefania Parisi, 'Byssus: Weaving of Sea Silk', *Fashion World*, October 19, 2014, accessed January 6, 2019, http://www.technofashionworld.com/byssus -weaving-of-sea-silk/.

136 **its diameter can vary:** On the physical properties of byssus, see 'Fibre Analysis of the Byssus of *Pinna nobilis* L.', Project Sea-silk, accessed January 6, 2019, http://www.muschelseide.ch/en /biologie/byssus/faseranalyse.html.

137 **'cities and villages in the depth of the sea':** Johann Chemnitz, 'Abhandlung von der Steckmuschel und ihrer Seide, wie auch vom Pinnenwächter', *Der Naturforscher* 10 (1777): 2 (trans. mine).

137 **their numbers began to decline:** On the threats facing the *Pinna nobilis*, see 'The Noble Pen Shell (*Pinna nobilis* L.)', Project Sea-silk, accessed January 6, 2019, http://www.muschelseide .ch/en/biologie/pinna-nobilis.html.

137 **In 1992 the Italian government declared the *Pinna nobilis*:** European Council Directive 92/43/EEC, *Pinna nobilis*.

137 **discovered a forest of some two hundred *Pinna nobilis*:** Fiona Ehlers and Christian Wüst, 'Salvaging the "Costa Concordia": Doomed Cruise Ship Prepares for Final Voyage', *Der Spiegel International Online*, September 26, 2012, accessed January 6, 2019, http://www.spiegel.de /international/europe/how-the-costa-concordia-will-be-salvaged-a-857683.html.

138 **'The Pinna is torn off the rocks with hooks':** Swinburne, *Travels in the Two Sicilies*, 79.

138 **'The old maestro leaves'**: 'Sant'Antioco, intervista a Vigo Chiara', *Sardegna Digital Library*, November 27, 2008, accessed January 6, 2019, http://www.sardegnadigitallibrary.it/index.php?xsl= 2436&id=198665 (transcription and trans. mine).

140 **'No more ploughs'**: Bachisio Bandinu and Salvatore Cubeddu, *Il quinto moro: Soru e il sorismo* (Selargius, Sardinia: Domus de Janas, 2007), 70 (trans. mine).

140 **'traditions or living expressions inherited'**: See 'What Is Intangible Cultural Heritage?' on the website of UNESCO, accessed January 6, 2019, https://ich.unesco.org/en/what-is -intangible-heritage-00003.

140 **Justinian I sent 'certain monks'**: On sericulture in Byzantium, see Heleanor B. Feltham, 'Justinian and the International Silk Trade', *Sino-Platonic Papers* 194 (November 2009).

141 **Giuseppe Capecelatro, the archbishop of Taranto**: On this key figure in the history of sea silk, see 'Giuseppe Capecelatro, Archbishop of Taranto', Project Sea-silk, accessed January 6, 2019, http://www.muschelseide.ch/en/geschichte/neuzeit/giuseppe-capecelatro.html.

141 **she reportedly wove**: See, for example, Carlotta Lombarda, 'Basilia, lo stemma e una "sacerdotessa"', *Corriere della Sera*, May 25, 2010.

141 **'Byssus cannot be bought or sold'**: Redazione *Fame di Sud*, 'Chiara Vigo'.

142 **'There is not a single sign'**: 'Sant'Antioco, intervista a Vigo Chiara' (transcription and trans. mine).

143 **were in fact fictions**: See McKinley, 'Pinna and Her Silken Beard', 9–223.

143 **linguistic error in the fifteenth century**: For a detailed treatment of Gaza's error, see Pieter van der Feen, 'Byssus', *Basteria* 13, no. 4 (1949): 66–71; and Felicitas Maeder, 'Irritating Byssus: Etymological Problems, Material Facts, and the Impact of Mass Media,' in *Textile Terminologies from the Orient to the Mediterranean and Europe, 1000 BC–1000 AD*, ed. Salvatore Gaspa, Cécile Michel and Marie-Louise Nosch (Lincoln, NE: Zea Books, 2017), 500–19.

144 **he'd dated it to the fourth or fifth century**: See Antonio Taramelli, 'Scoperte di antichità nell'antica Sulcis,' *Notizie degli scavi di antichità* 5 (1908): 151–52.

144 **a professor of Hebrew studies at the Sorbonne**: Although he was described as a professor of Hebrew codicology at the École Pratique des Hautes Études (EPHE) – see Paradiso, 'Chiara Vigo' – I was informed by the EPHE that Gabriel Hagai has 'neither any claim to the title of professor nor any professional affiliation to EPHE' (email dated June 1, 2016).

144 **'Since it cannot be loosened'**: Giuseppe Saverio Poli, *Testacea utriusque Siciliae, eorumque historia et anatome tabulis aeneis illustrata* (Parma: Ex Regio Typographeio, 1795), vol. 2, 231, quoted and translated in. Gilroy, *The History of Silk*, 181.

145 **trim the beards of thousands of shells**: Helen Scales, *Spirals in Time*, 164–66.

145 **plundering a selection of Chiara's quotations**: The sources of these quotations are given in these endnotes; quotations without corresponding endnotes are drawn from my own conversations with her.

146 **'the truth about byssus becomes little by little rejected'**: Claudio Moica, 'Si scoprono nuovi maestri della tessitura: Il bisso a Sant'Antioco', *Gazzetta del Sulcis-Iglesiente*, July 10, 2014 (trans. mine).

147 **Italo Diana started experimenting**: On the history of Italo Diana's studio, see Ginevra Zanetti, 'Un'antica industria sarda: Il tessuto d'arte per i paramenti sacri', *Archivio Storico Sardo* 29 (1964): 261; Sergio Flore, 'L'isola di Sardegna . . . l'ultimo filo del *panno del mare*', in *Bisso marino – Fili d'oro dal fondo del mare*, 59–60; Gerolama Carta Mantiglia, 'Il bisso marino in Sardegna', in *Bisso marino – Fili d'oro dal fondo del mare*, 53–54; Claudio Moica, 'Italo Diana: Il misterioso maestro del bisso di Sant'Antioco', *Gazzetta del Sulcis-Iglesiente*, July 31, 2014; and Claudio Moica, 'Italo Diana ricordato dai figli di Jolanda Sitzia: L'allieva e la rievocazione del maestro', *Gazzetta del Sulcis-Iglesiente*, October 9, 2014; Claudio Moica, 'Emma Diana racconta il padre Italo Diana maestro del bisso', *Gazzetta del Sulcis-Iglesiente*, April 16, 2015.

148 **a linen tapestry . . . covered in embroideries:** You can find images of the tapestry, before and after it was altered, in 'Inventory: Objects', Project Sea-silk, accessed January 6, 2019, http://www.muschelseide.ch/en/inventar/Objekte/wandteppich.html.

148 **dropped off a special guest:** To recount Mussolini's visit to Carbonia, I relied on footage from *Carbonia*, directed by Fernando Cerchio (1940); Paola Atzeni, 'Il discorso di Carbonia: "Logos e polis"', *La ricerca folklorica* 58 (October 2008): 121–36; Manlio Brigaglia, *Il carbone sbagliato*, RAI documentary, 1969; exhibition material at the Museo del Carbone in Carbonia; and Federico Caprotti, *Mussolini's Cities: Internal Colonialism in Italy, 1930–1939* (Youngstown, NY: Cambria Press, 2007).

150 **'In its name,' he told the inhabitants:** The quotations of Mussolini are from Atzeni, 'Il discorso di Carbonia' (trans. mine).

151 **But no fascist official supported . . . came to nothing:** On sea silk and fascism, see Rita del Bene, 'Tessuti di bisso: Lana-pinna o Lana pesce', *1ᵃ Mostra Jonica d'Arte Sacra, 2–31 maggio 1937-XV* (Taranto: A. Scrimieri, 1937), 1–7; and Lucia D'Ippolito, 'Fra antiche tradizioni e ambizioni industriali: La produzione di bisso marino a Taranto', in *Bisso marino – Fili d'oro dal fondo del mare*, 73–113.

152 **more and more people needed the dark fabric:** On orbace and fascism, see Eugenia Paulicelli, *Fashion Under Fascism: Beyond the Black Shirt* (Oxford: Berg, 2004), 108; and Perry Wilson, 'The Nation in Uniform? Fascist Italy, 1919–43', *Past and Present* 221, no. 1 (November 2013): 249–72; Pier Gavino Vacca, 'Dai vecchi telai al museo della lana, la storia dell'Alas: L'antico stabilimento sarà riadattato per iniziativa dell'amministrazione di Macomer', *La Nuova Sardegna*, May 21, 2002.

152 **'*Qui giace Starace*':** Simonetta Fiori, 'Mise l'Italia in orbace e finì sui muri: Starace chi legge', *La Repubblica*, July 26, 2000.

154 **'I learned it from my grandmother':** Unnamed author, 'Nessuno l'aiuta: Una lunga guerra contro la burocrazia', *La Nuova Sardegna*, February 6, 2000. Original text: '*L'ho imparata da mia nonna Maria Maddalena Mereu, che a sua volta era stata allieva di Italo Diana*' (trans. mine).

158 **'the true and real "discoverer"':** Metello Vene, 'Seguendo il filo di Chiara', *Airone* (supplement to no. 157), May 1994.

158 **Christian physician, Antiochus:** As one might expect, there are numerous popular variations on this story. See, for example, Saint Antiochus's entry in the online encyclopaedia of saints *Santi beati e testimoni*, accessed January 6, 2019, http://www.santiebeati.it/dettaglio/90511.

160 **passed from women to other women:** Claudio Moica, 'Efisia Murroni: L'ultima allieva sul bisso del poliedrico maestro Italo Diana', *Gazzetta del Sulcis-Iglesiente*, September 18, 2014; and 'Le sorelle Pes maestre di tessitura e di bisso: La passione di Assuntina e Giuseppina', *Gazzetta del Sulcis-Iglesiente*, October 23, 2014.

161 **wrote of 'trees which bore a fruit that became birds':** *The Travels of Sir John Mandeville*, trans. C. W. R. D. Moseley (Harmondsworth, UK: Penguin Books, 1983), 65, quoted in Florike Egmond and Peter Mason, 'Report on a Wild Goose Chase', *Journal of the History of Collections* 7, no. 1 (1995): 30.

161 **'nothing wanting, as to the external parts':** Sir Robert Moray, 'A Relation Concerning Barnacles', *Philosophical Transactions* 12 (1677): 926.

161 **'The "monks of old," and "the barefooted friars"':** Henry Lee, *Sea Fables Explained* (London: William Clowes and Sons, 1883), 122.

162 **threatened to take legal action against the Pes sisters:** Email from Chiara Vigo to the mayor of Sant'Antioco and other officials, April 26, 2013.

163 **certain academic publications continued:** See, for example, *I maestri del bisso, della seta, del lino: The Masters of Byssus, Silk and Linen*, ed. Małgorzata Biniecka (Rome: Sapienza Università Editrice, 2017), 3.

163 'The True History of Byssus in Sant'Antioco': 'La vera storia del bisso marino a Sant'Antioco', accessed January 6, 2019, https://www.facebook.com/groups/332760296881499/about/.

164 'They ruined my Christmas': Monica Magro, 'Chiara Vigo pronta ad andare via: "Porterò all"estero l'arte del bisso"', *Sardinia Post*, March 20, 2016, https://www.sardiniapost.it/cronaca /il-caso-chiara-vigo-pronta-a-lasciare-lisola-io-tradita-portero-allestero-larte-del-bisso/ (trans. mine).

164 Chiara promptly enlisted the support: ANSA staff, 'Flash mob per Museo Bisso a S. Antioco', ANSA, January 24, 2016, accessed January 6, 2019, http://www.ansa.it/sardegna/notizie/2016 /01/24/flash-mob-per-museo-bisso-a-s.antioco_b967f059-0b9b-4c01-b8a8-4fa5f3cd8522.html; unnamed author, 'Museo del Bisso, Pigliaru e Firino: "Necessario trovare una soluzione"', *Sardinia Post*, January 26, 2016, accessed January 6, 2019, https://www.sardiniapost.it/politica /museo-del-bisso-pigliaru-e-firino-necessario-trovare-una-soluzione/; Ilenia Mura, 'Sindaci alla corte del bisso', *L'Unione Sarda*, October 12, 2016.

165 Weaving has often been associated: See Kathryn Sullivan Kruger, *Weaving the Word: The Metaphysics of Weaving and Female Textile Production* (Selinsgrove, PA: Susquehanna University Press, 2001), 144.

165 'In a universe almost totally': 'Salviamo l'ultimo maestro di bisso: Chiara Vigo', petition to the governor of Sardinia, Francesco Pigliaru, accessed January 6, 2019, https://www.change.org /p/sardegna-salviamo-l-ultimo-maestro-di-bisso-marino-chiara-vigo-quirinalestampa-f-pigliaru -comunesantioco (trans. mine).

166 'Byssus today is Chiara': Fabrizio Steri, comment on 'Nessuno tocchi Il Museo del Bisso', Facebook, October 14, 2016, accessed January 6, 2019, https://www.facebook.com/groups /274249602779746/about/ (trans. mine).

167 'The few people remaining': Brigaglia, *Il carbone sbagliato* (trans. mine).

167 so-called Plan for Rebirth: See Russell King, 'Recent Industrialisation in Sardinia: Rebirth or Neo-Colonialism?', *Erdkunde* 31, no. 2 (June 1977): 99; Guy Dinmore and Giulia Segreti, 'Sun Sets on Sardinia's Mineral Industries', *Financial Times*, September 2, 2012; and Ugo Rossi, 'There's No Hope: The Global Economic Crisis and the Politics of Resistance in Southern Europe', *Belgeo* 1–2 (March 2012): 1–18.

168 special chemosynthetic bacteria: Sebastién Duperron et al., 'Symbioses Between Deep-Sea Mussels (Mytilidae: Bathymodiolinae) and Chemosynthetic Bacteria: Diversity, Function and Evolution', *Comptes Rendus: Biologies*, 332, no. 2–3 (February – March 2009): 298–310.

VICUÑA FIBRE

173 the Byssus Museum was closed: ANSA staff, 'Chiude tra le polemiche Museo del Bisso', ANSA, October 4, 2016, accessed January 6, 2019, http://www.ansa.it/sardegna/notizie/2016 /10/04/chiude-tra-le-polemiche-museo-del-bisso_499f0676-3f73-4810-83bf-9171eca9ff11.html; and 'Il Comune di Sant'Antioco ha sfrattato Chiara Vigo dal Montegranatico: Chiude il Museo del Bisso', *La Provincia del Sulcis Iglesiente*, October 1, 2016, accessed January 6, 2019, http:// www.laprovinciadelsulcisiglesiente.com/wordpress/2016/10/il-comune-di-santantioco -ha-sfrattato-chiara-vigo-dal-montegranatico-chiude-il-museo-del-bisso/.

173 sourced her byssus from a different species: See also Claudio Moica, 'Arianna Pintus, la jana del bisso', *Tottus in Pari*, March 11, 2018.

174 alarm about a mass mortality event: Maite Vázquez-Luis et al., 'S.O.S. *Pinna nobilis*: A Mass Mortality Event in Western Mediterranean Sea', *Frontiers in Marine Science*, July 17, 2017.

174 In 1958 the United States was gripped by a political scandal: James Brooke, 'Sherman Adams Is Dead at 87: Eisenhower Aide Left Under Fire', *The New York Times*, October 28, 1986; and 'The Adams Case', *Newsweek*, June 23, 1958.

175   **governors of forty-eight states:** *Investigation of Regulatory Commissions and Agencies: Hearings Before a Subcommittee of the Committee on Interstate and Foreign Commerce*, House of Representatives, 85th Congress, Second Session, part 10, June 17, 1958, 3717.

175   **'I don't know any other way':** Unnamed author, 'The Vicuña Man', *Newsweek*, October 2, 1967.

175   **'As I look back now':** Sherman Adams, *Firsthand Report: The Story of the Eisenhower Administration* (New York: Harper & Brothers, 1961), quoted in Brooke, 'Sherman Adams Is Dead at 87'.

175   **'vicuña man':** Unnamed author, 'The Vicuña Man'.

175   **diameter of 12.5 microns:** Jane Wheeler, 'South American Camelids – Past, Present and Future', *Journal of Camelid Science* 5 (2012): 6.

175   **all wore vicuña coats:** Meg Lukens Noonan, *The Coat Route: Craft, Luxury & Obsession on the Trail of a $50,000 Coat* (New York: Spiegel & Grau, 2013), 47.

175   **'When a mink wants to give a mink':** 'The Education of Sherman Adams', *The Philadelphia Inquirer*, June 22, 1958.

176   **They are in fact cousins of the two-humped camel:** For background on the vicuña, I relied on: Wheeler, 'South American Camelids;' Carl Koford, 'The Vicuña and the Puna', *Ecological Monographs* 27, no. 2 (April 1957), 153–219; William Franklin, 'High, Wild World of the Vicuña', *National Geographic* 143 (1973), 76–91.

176   **only the Incan emperor himself:** Garcilaso de la Vega, *The Royal Commentaries of Peru in Two Parts*, trans. Sir Paul Rycaut (London: Miles Flesher, 1688), vol. 1, 195.

177   **eighty thousand animals per year were killed:** Unknown 'early nineteenth century' author, cited in A. Cabrera and J. Yepes, *Mamíferos sud-americanos* (Buenos Aires: Compañía Argentina de Editores, 1940), cited in Koford, 'The Vicuña and the Puna', 213.

177   **'such havock hath been made':** Vega, *The Royal Commentaries*, vol. 1, 195, quoted in Noonan, *The Coat Route*, 45–46.

177   **an earlier political scandal:** In 1951 a Senate subcommittee uncovered that E. Merl Young, a former official of the federal Reconstruction Finance Corporation (RFC), had received a gift of a $9,450 mink coat from a Washington lawyer who worked for a firm that had received a loan from the RFC. See 'Merl Young of R.F.C.; Was Named in Scandal', *The New York Times*, August 22, 1981.

178   **numbered some ten thousand:** Jane Wheeler and Domingo Hoces, 'Community Participation, Sustainable Use, and Vicuña Conservation in Peru', *Mountain Research and Development* 17, no. 3 (August 1997): 284.

178   **Felipe Benavides, a Peruvian conservationist:** Joseph Novitski, 'Legislation by Bolivia and Peru Fails to Halt Threat to Vicuna', *The New York Times*, March 1, 1970.

178   **Since the 1930s he had investigated:** On Benavides's investigations and campaigns, see Wilfredo Pérez Ruiz, *La saga de la vicuña* (Lima: CONCYTEC, 1994), 49–50; Felipe Benavides, letter to the editor, *The Guardian*, November 9, 1967; 'Dwindling Vicuna', *South China Morning Post*, December 11, 1967; Dennis Barker, 'Vicuna Man Spots British Loophole', *The Guardian*, February 14, 1975; Marlene Simons, 'The Agitator of the Andes', *Los Angeles Times*, February 24, 1975; Karen DeYoung, 'Peruvian Wages Battle to Save Baby Seals', *Los Angeles Times*, June 21, 1978; Michael Reid, 'Obituary: Felipe Benavides, Saviour of the Vicuna', *The Guardian*, February 27, 1991; and Alex Emery, 'Peruvian Ecologist Felipe Benavides Dies in London', Associated Press, February 22, 1991.

178   **'The persecuted majority':** *Christian Science Monitor Dispatch*, 'Wildlife Champion Savior of People', Orlando, Florida, *Sentinel Star*, April 13, 1975.

178   **'tall elegant man of fifty-eight':** Faith McNulty, 'Peruvian Conservationist', *The New Yorker*, October 4, 1976.

179   **'I went around my country':** Ibid.

179   **'What is going on?':** Ibid.

179 the vicuña flourished in the park's early years: Wheeler and Hoces, 'Community Participation', 284.

179 the Smithsonian declared: Stephen Banker, 'And Now, Good News: Endangered Species Is Saved', Smithsonian, January 1977, 60–64.

179 A photograph of Benavides: Pérez, La saga de la vicuña, dedication.

179 three-to-five-year prison sentence: Novitski, 'Legislation by Bolivia and Peru'.

179 'looks best of all on vicuñas': Felipe Benavides, 'Vanishing Vicunas', letter to the editor, The Guardian, November 9, 1967.

179 bad drought in Pampas Galeras: Wheeler and Hoces, 'Community Participation', 284.

180 'The cheapest and most effective way': Brack in Lima Times, quoted in Jorge Castro de los Ríos, 'Masacre en la Puna', Caretas, June 4, 1979 (trans. mine).

180 'The meat is lean': Antonio Brack Egg, 'Historia del manejo de la vicuña en el Perú', Boletín de Lima 50 (March 1987): 74 (trans. mine).

180 New Zealand began culling its red deer population: Guil Figgins and Peter Holland, 'Red Deer in New Zealand: Game Animal, Economic Resource or Environmental Pest?', New Zealand Geographer 68 (2012): 36–48.

180 'It would be possible': Brack, 'Historia del manejo', 74 (trans. mine).

180 'waste country' into 'a very paradise for the deerstalker': R. A. Loughnan, New Zealand: Notes on Its Geography, Statistics, Land System, Scenery, Sport and the Maori Race (Wellington: J. Mackay, 1902), 28, quoted in Figgins and Holland, 'Red Deer in New Zealand', 38.

180 South America's great quadrupeds: Wheeler, 'South American Camelids', 2.

180 'To offer vicuña chops': 'La saca funesta', Caretas, November 17, 1986 (trans. mine).

181 Benavides had gone after anyone: Simons, 'The Agitator of the Andes'; and DeYoung, 'Peruvian Wages Battle'.

181 'Not even rabbits reproduce': Peter Gwynne and Larry Rohter, 'Open Season on Vicuñas', Newsweek, November 12, 1979.

181 there were but a third: Wheeler and Hoces, 'Community Participation', 284.

181 'They wound the animals': Smyth et al., 'Vicuna Hunt', The Observer, May 27, 1979.

181 more and more personal and embittered: Perhaps unsurprisingly, the sources on the dispute are polarized. For a critical take on the cull and Brack, see Castro de los Ríos, 'Masacre en la Puna'; 'El cuento de la vicuña', Caretas, May 5, 1980; Pablo Grimberg, 'Disparos sin reserva,' Caretas, July 14, 1980; 'Confirmando denuncias: Una comisión parlamentaria comprueba la debacle de la vicuña', Caretas, December 28, 1981; 'La saca funesta'; and Pérez, La saga de la vicuña. For a defence of the cull, see Brack, 'Historia del manejo'; and Marc Dourojeanni, 'Reserva Nacional de Pampa Galeras: La primera década', unpublished report, Derecho, Ambiente y Recursos Naturales (DAR), Lima Agenda, 2014. For a balanced view, see Wheeler and Hoces, 'Community Participation'.

181 an aloof aristocrat: This view still persists today in Pampas Galeras and the campesino community of Lucanas. Although the international press lionised Benavides as the 'saviour of the vicuña', I found his name to be toxic in rural Ayacucho.

182 Benavides won the war: Brack, 'Historia del manejo', 73.

182 But the dispute consumed: Emery, 'Peruvian Ecologist'; David Nicholson-Lord, 'World Wildlife Fund Negotiated Purchase of Guns to Hunt Rare Animal', The Independent on Sunday, January 13, 1991.

182 His obituary in El Expreso: Fernando Ramírez Alfaro, 'El día que las vicuñas lloraron', El Expreso, January 31, 1991, cited in Wilfredo Pérez Ruiz, 'Felipe Benavides: Un amigo del Partido del Pueblo', La Tribuna, June 24, 2005 (trans. mine).

182 'Those dogs of Tumbez and Puerto Viejo': Pedro Pizarro, Relation of the Discovery and Conquest of the Kingdoms of Peru, trans. Philip Ainsworth Means (New York: Cortes Society, 1921), vol. 1, 224. This episode is elegantly recounted by John Hemming, The Conquest of the Incas (New York: Harcourt Brace Jovanovich, 1970), 50–51.

182 'There was so much cloth of wool and cotton': John Murra, 'Cloth and Its Functions in the Inca State', *American Anthropologist* 64, no. 4 (August 1962): 717.

183 'I could not say about the warehouses': Pedro Pizarro, 'Relación del descubrimiento y conquista de los reinos del Perú', in *Colección de documentos inéditos para la historia de España* (Madrid, 1844), vol. 5, 272, quoted in translation in Murra, 'Cloth and Its Functions', 717.

183 'There is nothing strange': Murra, 'Cloth and Its Functions', 721.

183 'No political, military, social, or religious event': Ibid., 722.

183 The Jesuit Bernabé Cobo related: Bernabé Cobo, 'Historia del nuevo mundo', in *Biblioteca de autores españoles* (Madrid: Ediciones Atlas, 1956), vol. 91–92, 258–59, cited in Murra, 'Cloth and Its Functions', 711.

183 the Incas harvested vicuña fibre in a *chaku*: Among the chronicles there is great variation in the descriptions of the *chaku*. See Helen Cowie, *Llama* (London: Reaktion Books, 2017), 38–39; and Hugo Yacobaccio, 'The Historical Relationship Between People and the Vicuña', in *The Vicuña: The Theory and Practice of Community-Based Wildlife Management*, ed. Iain Gordon (New York: Springer, 2009), 11–12.

184 'Those great conservationists': Felipe Benavides, 'From the Incas to C.I.T.E.S.', paper presented at 3rd World Wilderness Congress, Scotland, October 8–15, 1983, 3. I am grateful to Wilfredo Pérez Ruiz for sharing this document.

184 'The people of the Andes': Simons, 'The Agitator of the Andes'.

184 shear tame vicuñas: Reid, 'Obituary: Felipe Benavides'.

184 'I want everybody to have': DeYoung, 'Peruvian Wages'; see also McNulty, 'Peruvian Conservationist'.

184 selected Peruvian vicuña populations: Wheeler and Hoces, 'Community Participation', 285. On Benavides's role, see Pérez, *La saga de la vicuña*.

184 when President Fujimori signed a decree: Wheeler and Hoces, 'Community Participation', 285–86.

185 'throttle . . . Peru's cities to death': Mario Vargas Llosa, *A Fish in the Water: A Memoir*, trans. Helen Lane (New York: Farrar, Straus and Giroux, 1994), 215.

186 little historical information concerning the Incan *chaku*: On the roots of the *chaku*, see Tetsuya Inamura, 'Chacu Collective Hunting of Camelids and Pastoralism in the Peruvian Andes', *Global Environmental Research* 10 (2006): 39–48.

186 'a set of practices, normally governed': Eric Hobsbawm, 'Introduction: Inventing Traditions', in *The Invention of Tradition*, ed. Eric Hobsbawm and Terence Ranger (Cambridge, UK: Cambridge University Press, 2004), 1.

186 hundreds of campesino communities: Wheeler and Hoces, 'Community Participation', 286.

187 the Sami faced similar problems: See Gudrun Norstedt, Anna-Maria Rautio and Lars Östlund, 'Fencing the Forest: Early Use of Barrier Fences in Sami Reindeer Husbandry', *Rangifer* 37, no. 1 (2017): 69–92.

187 CONACS, the state body: Consejo Nacional de Camélidos Sudamericanos (National Council for South American Camelids).

187 encouraging local communities to build wooden fences: On the use of fences, see Gabriela Lichtenstein et al., *Manejo comunitario de vicuñas en Perú: Estudio de caso del manejo comunitario de vida silvestre* (London: IIED, 2002).

188 Peru's vicuñas more than tripled: For figures on vicuña populations, see *Censo poblacional de vicuñas 2012*, Dirección General Forestal y de Fauna Silvestre (DGFFS); Gabriela Lichtenstein et al., 'Vicugna vicugna [Vicuña]', *The IUCN Red List of Threatened Species*, 2008, and P. Acebes et al., 'Vicugna vicugna' for the 2018 update.

188 'I trim [rhino] horn in such a way': Tony Carnie, 'Meet the World's Largest Rhino Breeder', *The Mercury*, September 6, 2016; Jani Actman, 'Can You Save Rhinos by Selling Their Horns?', *National Geographic*, August 20, 2017.

189 **During the 1960s and 1970s:** On the history of Pampas Galeras, see Wheeler and Hoces, 'Community Participation', 284.

191 **almost seventy thousand lives:** '¿Cuántos peruanos murieron? Estimación del total de víctimas causadas por el conflicto armado interno entre 1980 y el 2001', Informe Final de la CVR Anexo 2, Comisión de la Verdad y Reconciliación, August 28, 2003, 13.

191 **Andean myth that tells that the Incan emperor:** 'Q'olla myth', cited in Amy Cox, 'Politics of Conservation and Consumption: The Vicuña Trade in Peru' (master's thesis, University of Florida, 2003), 17. I have been unable to verify the origin of this story.

192 **the Sinchis, the government's death squads:** On the Sinchis, see Lewis Taylor, 'Counter-Insurgency Strategy, the PCP-Sendero Luminoso and the Civil War in Peru, 1980–1986', *Bulletin of Latin American Research* 17, no. 1 (1998): 35–58.

193 **The vicuña is built for altitude:** See Klaus D. Jürgens et al., 'Oxygen Binding Properties, Capillary Densities and Heart Weights in High Altitude Camelids', *Journal of Comparative Physiology B*, 158, no. 4 (1988): 469–77.

195 **'certain Indians entered the enclosure':** Pedro Cieza de León, *The Second Part of the Chronicle of Peru*, ed. and trans. Clements R. Markham (London: Hakluyt Society, 1883), 46.

195 **Shearing is an ancient technique:** Robert Jacobus Forbes, *Studies in Ancient Technology* (Leiden: E. J. Brill, 1987), vol. 4, 7.

197 **'It is thick and bushy':** Don Vincente Pazos, *Letters on the United Provinces of South America Addressed to the Hon. Henry Clay*, trans. Platt H. Crosby (New York: J. Seymour, 1819), 226.

198 **biologists have tried to understand:** On the effects of the *chaku*, see Catherine Sahley, Jorge Torres Vargas and Jesús Sánchez Valdivia, 'Biological Sustainability of Live Shearing of Vicuña in Peru', *Conservation Biology* 21, no. 1 (February 2007): 98–105; Cristian Bonacic, David Macdonald and Ruth Feber, 'Capture of the Vicuña (*Vicugna vicugna*) for Sustainable Use: Animal Welfare Implications', *Biological Conservation* 129 (2006): 543–50; Yanina Arzamendia and Cristian Bonacic, 'Behavioural and Physiological Consequences of Capture for Shearing of Vicuñas in Argentina', *Applied Animal Behaviour Science* 125 (2010): 163–70.

198 **In an interview with *National Geographic*:** Katarzyna Nowak, 'Legalizing Rhino Horn Won't Save Species, Ecologist Argues', *National Geographic*, January 8, 2015.

199 **'most curious case of selection':** Charles Darwin, *The Variation of Animals and Plants Under Domestication* (London: John Murray, 1888), vol. 2, 192.

200 **'[The animals] were afterwards fed':** M. F. de Theran, 'An Account of the First Attempt Made in Spain to Naturalise and Domesticate the Vicuna (Camelus Vicugna;) the Alpaco (Camelus Paco;) and the Lama (Camelus Glama;) to which Are Added, Some Observations on the Wool of These Animals', *London Journal of Arts and Sciences* 4, no. 12 (1822), 189. This article is a translation of Theran's 'Notice sur un premier essai fait en Espagne pour acclimater et réduire à l'état de domesticité les vigognes, les alpacos et les lamas, et sur la laine de ces animaux', in *Revue encyclopédique ou analyse raisoneé des productions les plus remarquables dans la literature, les sciences et les arts* 15, no. 43 (July 1822): 221–27. It is likely that 'Theran' is a French rendering of his actual name in Spanish, Terán.

200 **displacing flocks of camelids:** On the Spanish displacement of camelids, see Jane Wheeler, Angus Russel and Helen Stanley, 'A Measure of Loss: Prehispanic Llama and Alpaca Breeds', *Archivos de zootecnia* 41, no. 154 (extra) (1992), 467–75.

200 **as 'ovejas peruanas':** Daniel W. Gade, 'Llamas and Alpacas as "Sheep" in the Colonial Andes: Zoogeography Meets Eurocentrism', *Journal of Latin American Geography* 12, no. 2 (2013): 221–43.

201 **'In hatred of their late minister':** James Rennie, *The Menageries: Quadrupeds, Described and Drawn from Living Subjects* (London: Charles Knight, 1829), 325.

201 **a hat from camelid fibre:** Theran, 'An Account of the First Attempt', 192.

201 **the camelids perished within three years:** Juan Colón, cited in Antonio Cabral Chamorro, 'El jardín botánico Príncipe de la Paz de Sanlúcar de Barrameda: Una institución ilustrada al

servicio de la producción agraria y forestal', *Revista de estudios andaluces*, no. 21 (1995): 180. There are conflicting reports as to the fate of the camelids; see Cowie, *Llama*, 89–90.

201 **'After the experiment':** Theran, 'An Account of the First Attempt', 192–93.

202 **the 'Indians' of South America could domesticate vicuñas:** William Walton, *An Historical and Descriptive Account of the Peruvian Sheep, Called Carneros de la Tierra* (London: J. Harding, 1811), 137. Walton was dismissive of 'the Indian', writing that he or she 'had no idea of crossing the breed' (13) and 'is an enemy to innovation' (137).

202 **'The mountains of Wales':** Walton, *An Historical and Descriptive Account of the Peruvian Sheep*, 151.

202 **'It is a fact':** Ibid., 130.

202 **except by crossing them with alpacas:** Crosses between alpacas and vicuñas are known as paco-vicuñas and are larger and 'less gracile' than their wild progenitors. See Wheeler, 'South American Camelids', 14.

202 **'perpetual wildness':** This phrase is taken from Francis Galton's essay on domestication 'The First Steps Toward the Domestication of Animals', *Transactions of the Ethnological Society of London* 3 (1865): 122–38. On research concluding that the vicuña had never been domesticated, see Jane Wheeler, 'Evolution and Present Situation of the South American Camelidae', *Biological Journal of the Linnean Society* 54, no. 3 (March 1995): 281–84.

202 **proved to be deeply flawed:** Wheeler, 'Evolution and Present Situation,' 283–84.

203 **her findings were later confirmed through DNA:** Miranda Kadwell et al., 'Genetic Analysis Reveals the Wild Ancestors of the Llama and the Alpaca', *Proceedings: Biological Sciences* 268, no. 1485 (December 2001): 2575–84.

204 **critically endangered in Peru:** Ricardo Baldi et al., '*Lama guanicoe* [Guanaco]', *The IUCN Red List of Threatened Species*, 2016. Today a number of ecologists believe that the *chaku*, for all its compromises, is the 'best available option' for the vicuña. Email from Gabriela Lichtenstein, December 12, 2018; Dr Lichtenstein, an Argentine ecologist, has spent decades researching different *chakus* across South America.

205 **$200 worth of fibre becomes $10,000:** According to Alexander Kasterine and Gabriela Lichtenstein, the raw fibre used for a Loro Piana stole constitutes 2 per cent of its final value. See 'Trade in Vicuña: The Implications for Conservation and Rural Livelihoods', International Trade Centre, Geneva, 2018.

205 **a vicuña parka for $26,495:** 'Parka Vicuña: Vicuña Storm System', Loro Piana, accessed October 31, 2018, https://us.loropiana.com/en/p/Men/Vicuña/Outerwear-Jackets/Parka-Vicuña -FAG4074.

206 **The inhabitants of Lucanas identify with the Lucanas or Rucanas people:** For historical background on Lucanas, I relied on Sarah Jane Abraham, 'Provincial Life in the Inca Empire: Continuity and Change at Pulapuco, Peru' (PhD diss., University of California, Santa Barbara, 2010).

208 **between around $200 and $500 per kilo:** Prices for fibre have fluctuated wildly since the start of the *chaku*. See Gabriela Lichtenstein, 'Vicuña Conservation and Poverty Alleviation? Andean Communities and International Fibre Markets', *International Journal of the Commons* 4, no. 1 (February 2010): 108.

209 **an '*espejismo*,' a mirage:** On the question of vicuña fibre's profitability for local communities, see Lichtenstein et al., 'Manejo comunitario de vicuñas en Perú'; Lichtenstein, 'Vicuña Conservation and Poverty Alleviation?'; Kasterine and Lichtenstein, 'Trade in Vicuña'; and Oscar Franco, 'Un espejismo llamado vicuña', *La Revista Agraria* 137 (February 2012).

209 **the sale of fibre was managed:** For background on the fibre market, I relied on Catherine Sahley, Jorge Torres Vargas and Jesús Sánchez Valdivia, 'Neoliberalism Meets Pre-Columbian Tradition: Campesino Communities and Vicuña Management in Andean Peru', *Culture & Agriculture* 26, no. 1–2 (March 2004): 60–68; and Cox, 'Politics of Conservation and Consumption'.

209   **much of the altiplano and puna has remained isolated:** For background on the altiplano, I relied on Nils Jacobsen, *Mirages of Transition: The Peruvian Altiplano, 1780–1930* (Berkeley: University of California Press, 1993). Some 250 miles north-west of Lake Titicaca, Lucanas is not situated on the altiplano, but the community's remoteness and altitude mean that it faces challenges similar to those on the altiplano proper.

211   **The free market had triumphed in the vicuña trade:** See Sahley, Torres and Sánchez, 'Neoliberalism Meets Pre-Columbian Tradition'.

212   **'The compulsory issue of culturally valued commodities':** Murra, 'Cloth and Its Functions', 721.

212   **a garb for the advent of neoliberalism:** I am indebted to the work of Sahley, Torres and Sánchez, 'Neoliberalism Meets Pre-Columbian Tradition', for this observation.

TAGUA

217   **it is illegal to remove raw fibre:** Email from Alfonso Martínez, November 26, 2018.

217   **lifted a moratorium on the domestic trade:** Rachael Bale, 'Breaking: Rhino Horn Trade to Return in South Africa', *National Geographic*, April 5, 2017; and Russell Goldman, 'South African Court Ends Ban on Sale of Rhinoceros Horns', *The New York Times*, April 5, 2017.

217   **'No longer will rhino need':** Tony Carnie, 'Rhino Baron Shifts Blame for "Disappointing" First Horn Auction', *Sunday Times* (South Africa), August 26, 2017.

218   **'the greatest curiosity of the age':** Ebenezer Bowman, 'The Greatest Curiosity of the Age', *The Portland Directory and Reference Book* (Portland, ME: Brown Thurston, between 1852 and 1863).

218   **Fourier Transform Infrared Spectroscopy . . . is the most effective test:** Edgard O. Espinoza and Mary-Jacque Mann, *Identification Guide for Ivory and Ivory Substitutes* (Baltimore: WWF, 1992), 24.

218   **nut from a palm – *Phytelephas*:** On the natural history of tagua, see Misael Acosta Solís, 'Tagua or Vegetable Ivory: A Forest Product of Ecuador', *Economic Botany* 2, no. 1 (January – March 1948): 46–57; and Anders S. Barfod, 'A Monographic Study of the Subfamily Phytelephantoideae (Arecaceae)', *Opera Botanica* 105 (1991): 1–73.

219   **almost as dense as copper:** According to Wayne P. Armstrong, a botanist, tagua measures about 2.5 on the Mohs scale of mineral hardness, while a copper coin measures 3.5. See 'Vegetable Ivory: Saving Elephants & the Rain Forest', Wayne P. Armstrong, July 12, 2010, accessed January 6, 2019, https://www2.palomar.edu/users/warmstrong/pljan99.htm.

219   **'in form and colour the miniature head of a Negro':** Edward Albes, 'Tagua – Vegetable Ivory', *Bulletin of the Pan American Union* 37, no. 2 (August 1913): 194.

219   **The tagua crop . . . became an important part:** On tagua as a commodity, see Acosta Solís, 'Tagua or Vegetable Ivory'; Albes, 'Tagua – Vegetable Ivory'; Tatiana Hidrovo Quiñónez, *Historia de Manta en la región de Manabí* (Manta and Quito: Editorial Mar Abierto and Eskeletra Editorial, 2005), vol. 2, 202–4; Anders Barfod, 'The Rise and Fall of Vegetable Ivory', *Principes* 53, no. 4 (1989): 181–90; Anders Barfod, Birgitte Bergmann and Henrik Pedersen, 'The Vegetable Ivory Industry: Surviving and Doing Well in Ecuador', *Economic Botany*, 44, no. 3 (1990): 293–300. I also made cautious use of Klaus Calderón's history of vegetable ivory on the Corozo Buttons website, accessed January 6, 2019, http://www.corozobuttons.com/2014/07 /evolution-corozo-buttons-tagua-nuts/.

220   **'Every tusk, piece and scrap of ivory':** Henry Morton Stanley, *In Darkest Africa* (New York: Charles Scribner's Sons, 1891), vol. 1, 240.

220   **might expose bullets or deformities:** In an intriguing, but depressing, article from 1910, John Bland-Sutton, a British surgeon, related how objects can be found in ivory tusks. 'Ivory-turners', he writes, 'have known for more than a century that hard bodies, such as bullets and

spear-heads, are occasionally found embedded in the solid parts of elephants' tusks without any sign existing on the surface of the tusk to indicate the point of entry.' According to Bland-Sutton, the Royal College of Surgeons once held a billiard ball in which was found a bullet. See 'The Diseases of Elephants' Tusks in Relation to Billiard Balls', *The Lancet* 176, no. 4552 (November 1910): 1534–37.

220 **in the New York area more than twenty-five factories:** On the history of button production in New York state, see Jeff Ludwig, 'Retrofitting Rochester: Rochester Button Company', Roches-ter, New York, *Democrat & Chronicle* (in partnership with the Office of the City Historian), September 2, 2013; Paul Grebinger, 'The Button: Not a Simple Notion', in *Reflecting on Amer-ica: Anthropological Views of U.S. Culture*, ed. Clare L. Boulanger (New York: Routledge, 2016), 41–50; Acosta Solís, 'Tagua or Vegetable Ivory', 52–53.

220 **'It does not matter to the negro':** Acosta Solís, 'Tagua or Vegetable Ivory', 57. The view of Acosta Solís, the one-time director of Ecuador's Institute of Natural Sciences, on the reckless-ness of *'montuvios'* reflects wider divisions between the country's urban elites and its coastal populations. See Daniel Bauer, 'Identities on the Periphery: *Mestizaje* in the Lowlands of South America', *Delaware Review of Latin American Studies* 15, no. 2 (December 2014).

221 **turned to banana plantations:** Barfod, 'The Rise and Fall of Vegetable Ivory', 188.

221 **clear much of its forests:** Ibid., 188.

221 **'We divorced ourselves from the materials':** Norman Mailer, 'The Big Bite', *Esquire*, April 1963, in Norman Mailer, *The Presidential Papers* (New York: G. P. Putnam's Sons, 1963), 159, quoted in Jeffrey Meikle, *American Plastic: A Cultural History* (New Brunswick, NJ: Rutgers University Press, 1995), 244.

222 **Some three million years ago:** For background on South America's megafauna, I relied on Donald R. Prothero, *The Princeton Field Guide to Prehistoric Mammals* (Princeton, NJ: Prince-ton University Press, 2017).

223 **as Thor Hanson points out:** Thor Hanson, *The Triumph of Seeds: How Grains, Nuts, Kernels, Pulses, & Pips Conquered the Plant Kingdom and Shaped Human History* (New York: Basic Books, 2015), 19–52.

223 **swirling patterns:** The plant biologists Allan Witztum and Randy Wayne examined an old tagua button under a microscope and captured its cellular architecture in a series of beautiful images. See 'Button Botany: Plasmodesmata in Vegetable Ivory', *Protoplasma* 249, no. 3 (July 2012): 721–24.

224 **part of a surge in interest:** The interest is usually dated to the publication of a short paper by Charles Peters, Alwyn Gentry and Robert Mendelsohn in which they concluded that the value of NTFPs in a one-hectare plot in the Peruvian Amazon was greater than that of its timber. See 'Valuation of an Amazonian Rainforest', *Nature* 339 (June 1989): 655–56.

225 **Conservation International, a non-governmental organistion:** Laura Tangley, *The Tagua Initiative: Marketing Biodiversity Products* (Washington, DC: Conservation International, 1993), 6. See also Susan Katz Miller, 'Palm Nuts to Help Crack Poverty Problem' ,*New Scientist*, November 13, 1993.

225 **Firms like Patagonia:** '2 Companies Agree to Use Product from Rain Forest', *The New York Times*, September 12, 1990.

225 **According to the World Wildlife Fund:** See 'Habitat Loss', World Wildlife Fund, accessed November 1, 2018, http://wwf.panda.org/our_work/wildlife/problems/habitat_loss_degradation/.

225 **the palms could house or nourish many species:** Grischa Brokamp et al., 'Productivity and Management of *Phytelephas aequatorialis* (Arecaceae) in Ecuador', *Annals of Applied Biology* 164 (2014): 259.

226 **Manta first sought to fashion itself:** On the history of Manta, see DeWight R. Middleton, 'Migration and Urbanization in Ecuador: A View from the Coast', *Urban Anthropology* 8, no. 3–4 (Winter 1979): 313–32; and 'Development and Multiple Use: Conflict on an Ecuadorian Beach', *Urban Anthropology and Studies of Cultural Systems and World Economic*

Development 17, no. 4 (Winter 1988): 351–64. On Manta's tuna industry, see Nathan H. Bellinger, 'Globalization and Neoliberalism in Ecuador: The Expansion and Effects of the Commercial Tuna Fishing Industry' (master's thesis, University of Oregon, 2011). For figures on the tuna industry, see *Report of the Ecuadorian Tuna Sector*, Ministerio del Comercio Exterior, August 2017; Mónica Mendoza, 'Ecuador es el segundo productor de atún en el mundo, después de Tailandia', *El Comercio*, April 2, 2018.

227 **only 20 per cent of the tuna stays in Ecuador:** Mendoza, 'Ecuador es el segundo productor de atún'.

229 **'that species of *Phytelephas* were ivory bearing plants':** Barfod, 'A Monographic Study of the Subfamily Phytelephantoideae (Arecaceae)', 8.

229 **The botanical family of Arecaceae, or palm trees:** For background on palms, I relied on Grischa Brokamp, *Relevance and Sustainability of Wild Plant Collection in NW South America: Insights from the Plant Families Arecaceae and Krameriaceae* (Wiesbaden, Germany: Springer Spektrum, 2015).

230 **'in such a way that the blood':** Edward Grim, 'The Murder of Beckett' (1870), in *English Historical Documents*, ed. David C. Douglas, vol. 2, 768, quoted in *Portraits and Documents: England in the Early Middle Ages*, ed. Derek Baker (Dallas: Academia, 1993), 200.

231 **they grow incredibly slowly:** Barfod, 'A Monographic Study of the Subfamily Phytelephantoideae (Arecaceae)', 31–32. Acosta Solís writes that 'even at 14 or 15 years, the age at which it begins to flower and to produce fruit, the bases of the leaves are still in the earth, giving the appearance that the fruit emerges from the ground like a giant tuber.' See 'Tagua or Vegetable Ivory', 49.

231 **fretted that tagua seeds would be smuggled:** Acosta Solís, 'Tagua or Vegetable Ivory', 57.

232 **When you harvest tagua seeds:** On the question of sustainability of tagua harvesting, see Julie Velásquez Runk, 'Productivity and Sustainability of a Vegetable Ivory Palm (*Phytelephas aequatorialis*, Arecaceae) Under Three Management Regimes in Northwestern Ecuador', *Economic Botany* 52, no. 2 (April – June 1998): 168–82; and Brokamp et al., 'Productivity and Management'.

233 **treat the merits of commodification of forests with greater caution:** There is substantial literature that provides a critique of NTFP harvesting. See, for example, John Browder, 'The Limits of Extractivism', *BioScience* 42, no. 3 (March 1992): 174–82; and Douglas Southgate, Marc Coles-Ritchie and Pablo Salazar-Canelos, 'Can Tropical Forests Be Saved by Harvesting Non-Timber Products?', CSERGE Working Paper GEC 96-02 (1996). For a good summary of research on NTFPs, see Roderick P. Neumann and Erik Hirsch, 'Commercialisation of Non-Timber Forest Products: Review and Analysis of Research' (Bogor, Indonesia: Center for International Forestry Research, 2000).

233 **Certainly this has been proven:** Douglas Southgate, *Tropical Forest Conservation: An Economic Assessment of the Alternatives in Latin America* (New York: Oxford University Press), 1998, 56. See Brokamp et al., 'Productivity and Management', 258–59.

234 **I'd watched footage:** See 'Rhino Sedation and Horn Removal', *Trophy*, directed by Christina Clusiau and Shaul Schwarz, 2017.

234 **After decades of decline:** On the decline of button manufacturing in New York state, see Ludwig, 'Retrofitting Rochester'; and Grebinger, 'The Button', 41–50.

234 **town in southern Zhejiang province, China, named Qiaotou:** For background on Qiaotou, I relied on Jonathan Watts, 'The Tiger's Teeth', *The Guardian*, May 24, 2005; Seth Doane, 'A Look at China's "Button Town"', CBS News, October 8, 2015, accessed February 7, 2019, https://www.cbsnews.com/news/welcome-to-button-town-china/; and Rajah Rasiah, Xin-Xin Kong and Jebamalai Vinanchiarachi, 'Moving Up in the Global Value Chain in Button Manufacturing in China', *Asia Pacific Business Review* 17, no. 2 (2011): 161–74.

236 **difficulties of working with leather:** 'Some Notes on The Highfield Tanning Co., Runcorn' and 'Notes on Leather Tanning, History and Method', *Runcorn Historical Society*, c. 1967;

Katherine Norbury, *The Fish Ladder: A Journey Upstream* (London: Bloomsbury, 2015), 218–19.

236 **85 to 95 per cent of the original seed:** Email from Klaus Calderón, November 16, 2018.

238 **hoped to find employment in an oil refinery:** As of the time of writing, the Pacific Refinery shows little sign of materialising, despite repeated promises.

238 **may cut their threads if they are tightly sewn:** Email from Klaus Calderón, November 26, 2018.

238 **'a significant cost in dollars':** Harvey Hartman and Erika J. Haas, 'Patagonia Struggles to Reduce Its Impact on the Environment', *Total Quality Environmental Management* 5, no. 1 (Autumn 1995): 1–7.

239 **a documentary about tagua:** *Nuez de marfil*, directed by Florencia Luna, date unknown.

GUANO

245 **Reading up on *eiderdown*'s etymology:** *Oxford English Dictionary Online*, s.v. 'eider-down,' accessed January 5, 2019, http://www.oed.com/view/Entry/59968?redirectedFrom=eider-down&.

245 **'they have joined the eiderdown gang':** Stan Orme, Parl. Deb. (1971), vol. 811, col. 1616.

246 **the house I found myself in:** On the history of Tyntesfield, see James Miller, *Fertile Fortune: The Story of Tyntesfield* (London: National Trust, 2003).

246 **design on a stained-glass window:** I am indebted to Lesley Kinsley for pointing out this clue. See her articles 'From Textile to Guano Merchants: Antony Gibbs & Sons and Their Coastal Trade Links', National Trust *Views* 52 (Autumn 2015): 80; 'Guano, Science and Victorian High Farming: An Agro-Ecological Perspective', in *Victorian Sustainability in Literature and Culture*, ed. Wendy Perkins (Abingdon, UK: Routledge, 2018).

247 **the actual source of Gibbs's tremendous wealth:** For background on guano, I relied heavily on the work of Gregory T. Cushman: 'The Lords of Guano: Science and the Management of Peru's Marine Environment, 1800–1973' (PhD diss., University of Texas at Austin, 2003); and *Guano and the Opening of the Pacific World: A Global Ecological History* (Cambridge, UK: Cambridge University Press, 2013).

247 **Gibbs was unsure of the soundness:** Gibbs described the Lima branch's activities as an 'act of insanity'. See William M. Mathew, *The House of Gibbs and the Peruvian Guano Monopoly* (London: Royal Historical Society, 1981), 39.

247 **the demands of growing urban populations:** On soil exhaustion and its consequences, see Brett Clark and John Bellamy Foster, 'Guano: The Global Metabolic Rift and the Fertilizer Trade,' in *Ecology and Power: Struggles over Land and Material Resources in the Past, Present, and Future*, ed. Alf Hornborg, Brett Clark, and Kenneth Hermele (Abingdon, UK: Routledge, 2012), 68–82.

247 **12.7 million tonnes of guano:** Cushman, 'The Lords of Guano', 62.

247 **'The sunless lands of Great Britain':** 'The Guano Trade', *The Farmer's Magazine*, February 1855.

248 **the United States Congress passed the Guano Islands Act:** See Cushman, 'The Lords of Guano', 68–69.

248 **'I don't think much of the new discoveries':** Henry Gibbs to William Gibbs, February 9, 1855, *Antony Gibbs & Sons, Ltd. Copybook of in-letters addressed to William Gibbs, 1854–1855 London*, quoted in Mathew, *The House of Gibbs*, 176.

248 **'So great is the value of this branch':** Manuel Ortiz de Zevallos, 1858, quoted in Mathew, *The House of Gibbs*, 1.

248 **William received between 50 and almost 70 per cent:** Miller, *Fertile Fortune*, 34.

248 'The House of Gibbs that made their dibs': Lord Aldenham, quoted in Mathew, *The House of Gibbs*, 226. The Lord Aldenham in question is Antony Durant Gibbs, who was the fifth member of the Gibbs family to hold the title.

250 **some eleven stories high:** The American traveller George Washington Peck, who visited the Chinchas in 1853, described the guano on Chincha Norte as 150 feet deep; see *Melbourne and the Chincha Islands* (New York: Charles Scribner, 1854), 198. On the basis of old photographs, George Evelyn Hutchinson estimated that the size of the guano deposit on Chincha Central was 47.4 metres (155.5 feet). See George Kubler, 'Towards Absolute Time: Guano Archaeology', *Memoirs of the Society for American Anthropology* 4 (1948): 30.

250 **burying itself and anything it touches:** For evocative descriptions of guano, see Peck, *Melbourne and the Chincha Islands*, 169, 197–99. George Kubler details some of the objects found within Peru's guano deposits in 'Towards Absolute Time', 30.

251 **believed to hold enough excrement:** Wrote one visitor to the Chinchas in 1847: 'There is sufficient, at fifty thousand tons exported per annum, to last upwards of a thousand years!' Geo. Peacock, 'Stores of Guano on the Chincha Islands', *Nautical Magazine and Naval Chronicle*, March 1847. As Cushman reports, initial figures for the quantity of guano on the Chinchas, as high as 117 million tonnes, were wildly inaccurate; see 'The Lords of Guano', 73–74.

251 **surveyed, mined and sold:** For a detailed look at the means of guano extraction on the islands, see William Mathew, 'A Primitive Export Sector: Guano Production in Mid-Nineteenth-Century Peru', *Journal of Latin American Studies* 9, no. 1 (May 1977): 35–57.

252 **eight hundred Chinese on the Chinchas alone:** Mathew, 'A Primitive Export Sector', 41.

252 **toiling in the equatorial sun:** For working conditions on the islands, see Mathew, 'A Primitive Export Sector'.

252 **'That they are worked to death':** Peck, *Melbourne and the Chincha Islands*, 215.

252 **'Once on the islands':** Alanson Nash, quoted in Clark and Foster, 'The Global Metabolic Rift and the Fertilizer Trade', 78.

252 **According to the historian Watt Stewart:** Watt Stewart, *Chinese Bondage in Peru* (Durham, NC: Duke University Press, 1951). I relied on Mathew's summary of Stewart's work in 'A Primitive Export Sector', 42–43.

252 **some ninety-two thousand Chinese indentured labourers:** Cushman, 'The Lords of Guano', 71.

252 **the merchants who traded guano were largely distanced:** Mathew, 'A Primitive Export Sector', 36–37. Gibbs and Co. did briefly assume responsibility for the loading of guano in 1854, replacing Domingo Elías. According to Mathew, during this period wages were increased for labourers and there was a reported improvement in living conditions (45–46). Nevertheless, reports of appalling working conditions continued throughout the 1850s (see, for example, 'The Chincha Islands', *Nautical Magazine and Naval Chronicle*, April 1856).

252 **'If ever a philosopher's stone':** 'Guano,' *The Farmer's Magazine*, January – June 1854.

253 **'seem to me to be a kind of human abattoir':** Peck, *Melbourne and the Chincha Islands*, 204.

254 **'these same islands looked like creatures':** Alexander James Duffield, *Peru in the Guano Age* (London: Richard Bentley and Son, 1877), 89.

254 **marketing Chilean nitrates instead:** Kinsley, 'From Textile to Guano Merchants', 81.

254 **an American photographer, Henry Moulton:** On Moulton's career, see Keith McElroy, 'Henry De Witt Moulton: Rays of Sunlight from South America', *History of Photography*, 8, no. 1 (1984): 7–21.

254 **the Spanish had occupied the islands:** On the occupation of the guano islands, see Cushman, 'Guano and the Opening of the Pacific World', 56–57.

254 **various forms, shadows, and textures:** Moulton's extraordinary images were collated and published by Alexander Gardner in *Rays of Sunlight from South America* (Washington, DC: Philp

& Solomons, c. 1865). They can be viewed online on the website of the Library of Congress, at https://www.loc.gov/resource/rbc0001.2012gen13000/?st=gallery.

255 **employing the services of John Norton:** See Miller, *Fertile Fortune*, 59–88.

255 **Gibbs commissioned Lawrence Macdonald:** Ibid., 73–75.

255 **In 1872, Gibbs asked Arthur Blomfield:** Ibid., 99–107.

255 **'The beautiful house':** Ethel Romanes, *Charlotte Mary Yonge: An Appreciation* (Oxford: A. R. Mowbray, 1908), 158, quoted in Miller, *Fertile Fortune*, 98–99.

255 **acquired by the National Trust for £24 million:** Will Bennett, '"Magical" Victorian Mansion Saved for Nation in £24m Deal', *The Daily Telegraph*, June 19, 2002.

257 **'accumulation of bodies of animals and plants':** Such were the views, for example, of Arthur Mead Edwards, a New York chemistry professor. For his, and others', arguments that guano was not bird excrement, see May 1, 1871, meeting notes, *Proceedings of the Lyceum of Natural History in the City of New York*, ser. 1, 224–34 (233 for Edwards's phrase).

257 **kept as pets:** Peck, *Melbourne and the Chincha Islands*, 178–79.

258 **ornithologist named Robert Coker:** On Coker's career and work on the guano islands, see Cushman, 'The Lords of Guano', 128–37.

258 **he travelled out to the islands:** Robert E. Coker, 'Habits and Economic Relations of the Guano Birds of Peru,' *Proceedings of the U.S. Natural History Museum* 56 (1919): 449–50.

258 **'Unless one is enslaved to the freshwater bath':** Coker, 'Peru's Wealth-Producing Birds: Vast Riches in the Guano Deposits of Cormorants, Pelicans, and Petrels Which Nest on Their Barren, Rainless Coast', *National Geographic*, June 1920, 546.

258 **'If one awaits motionless':** Coker, 'Peru's Wealth-Producing Birds', 561.

258 **180,000 pairs of guanay cormorants:** Robert Coker, 'The Fisheries and the Guano Industry of Peru', *Bulletin of the Bureau of Fisheries* 28, no. 1 (1910), 359.

258 **'The fowl which dwelt':** Coker, 'Habits and Economic Relations', 484–85.

259 **the guano-producing birds were protected:** Ibid., 506.

259 **'innocent agents in the production':** Coker, 'Peru's Wealth-Producing Birds', 543.

259 **'Had Humboldt given correct information':** Coker, 'Habits and Economic Relations', 506.

259 **'It is necessary':** Robert E. Coker, 'Regarding the Future of the Guano Industry and the Guano-Producing Birds of Peru', *Science* 28, no. 706 (July 1908): 59.

260 **Compañía Administradora del Guano (CAG):** On the history of the CAG, see Cushman, 'Lords of Guano'.

260 **Rising precipitously some eighty metres:** For general background on Mazorca, I relied on Judith Figueroa et al., 'Línea Base Biológica de la Reserva Nacional Sistema de Islas, Islotes y Puntas Guaneras: Punta Salinas, Islas Huampanú y Mazorca (Lima)' (Lima: SERNANP, 2017).

261 **'No one is taken there by force':** Sebastião Salgado, *Workers: An Archaeology of the Industrial Age* (New York: Aperture, 1993), 16.

262 **'There is no moss, or lichen, or grass':** Duffield, *Peru in the Guano Age*, 91.

263 **'We experience a high rate of desertions':** Miguel Ángel Márquez, quoted in José Vadillo Vila, 'Potencia de la islas', *El Peruano*, July 20, 2016 (trans. mine).

264 **In the early years it was common:** Cushman, 'Lords of Guano', 171–72.

264 **the German scientists Fritz Haber and Carl Bosch:** On the Haber-Bosch Process and its relationship with guano, see Cushman, 'Lords of Guano', 159–63.

265 **1.2 per cent of all anthropogenic greenhouse gas emissions:** Sam Wood and Annette Cowie, *A Review of Greenhouse Gas Emission Factors for Fertiliser Production*, IEA Bioenergy, Paris, 2004, cited in Jeremy Woods et al., 'Energy and the Food System', *Philosophical Transactions of the Royal Society B: Biological Sciences* 365, no. 1554 (September 2010), 2997.

265 **Adolf Winter built a small wooden platform:** Winter's remarkable story is documented in Anita C. Gossow, *Die Guanoinsel auf Bird Rock: Das Lebenswerk Adolf Winters* (Windhoek: Namibia Wissenschaftliche Gesellschaft, 1995).

265 **As Gregory T. Cushman points out:** 'Lords of Guano', 161.

266 **the Spanish word for 'pelican,' *alcatraz*:** In addition to the Arabic *al-ġattās*, the *Oxford English Dictionary* (OED) suggests a possible association of *alcatraz* with *al-qadūs*, Arabic for 'a water bucket of a noria [water wheel]'. Given the pelican's voluminous bill, this seems fitting, even though pelicans do not carry water in their bills. See the OED entry for *alcatras* (a word once used in English to denote a pelican): *Oxford English Dictionary Online*, s.v. 'alcatras', accessed January 4, 2019, http://www.oed.com/view/Entry/4673?redirectedFrom=alcatraz. By contrast, the entry for *alcatraz* in the dictionary of the Real Academia Española is less charming. It suggests that *alcatraz* is possibly derived from the Andalusian Arabic *qaṭrás*, meaning 'el de andares ufanos', or 'a boastful or smug person' (trans. mine). See *Diccionario de la lengua española*, s.v. 'alcatraz', accessed January 4, 2019, http://dle.rae.es/?id=1dBGlf9|1dC3cpC. I prefer *al-ġattās*.

266 **conserving energy because of slight changes in airflow patterns:** See Paul R. Ehrlich, David S. Dobkin and Darryl Wheye, *The Birder's Handbook: A Field Guide to the Natural History of North American Birds* (New York: Simon & Schuster, 1988), 195.

266 **intrigued the ornithologist Robert Cushman Murphy:** Robert Cushman Murphy, *Bird Islands of Peru: The Record of a Sojourn on the West Coast* (New York and London: G. P. Putnam's Sons, 1925), 75–76.

267 **'I have seen them in double millions':** Duffield, *Peru in the Guano Age*, 93–94.

267 **was once highly proactive:** See Cushman, 'Lords of Guano', 174–78.

268 **the novel's protagonist, Giovanni Drogo, at first seems to be a medieval knight:** Tim Parks, 'Throwing Down a Gauntlet', *The Threepenny Review*, Winter 2001.

268 **'It often occurred to me':** Dino Buzzati, quoted in translation in Parks, 'Throwing Down a Gauntlet'.

270 **Countless purse seiners patrol:** On Peru's fishing industry and its ecological consequences, see Cushman, 'Lords of Guano', 430–89.

270 **'Nature does not allow animals':** Robert Cushman Murphy, 'Peru Profits from Sea Fowl', *National Geographic*, March 1959, 413.

271 **will continue to face great pressure:** For a detailed look at the threats facing the guanay, see BirdLife International 2018, '*Leucocarbo bougainvilliorum* [Guanay cormorant]', *The IUCN Red List of Threatened Species*, 2018.

271 **El Niño events will increase in frequency:** Axel Timmermann et al., 'Increased El Niño Frequency in a Climate Model Forced by Future Greenhouse Warming', *Nature* 398 (April 1999): 694–96.

EPILOGUE

275 **giant tunnel-boring machines:** Jon Excell, 'Crossrail: The Monster Tunnelling Under London Streets', BBC Future, June 3, 2015, accessed January 7, 2019, http://www.bbc.com/future/story/20150602-crossrail-the-monster-tunnelling-under-london-streets; and 'Meet Our Giant Tunnelling Machines', Crossrail, accessed January 7, 2019, http://www.crossrail.co.uk/construction/tunnelling/meet-our-giant-tunnelling-machines/.

275 **unearthed and classified:** Josephine McDermott, 'The Archaeological Legacy of the Crossrail Excavations', BBC News, February 10, 2017, accessed January 7, 2019, https://www.bbc.com/news/uk-england-london-38919314.

276 **'Though it might be nice to imagine':** Elizabeth Kolbert, *The Sixth Extinction: An Unnatural History* (New York: Picador, 2015), 235.

277 **'lesson that nature and culture':** Michael Pollan, *Second Nature: A Gardener's Education* (New York: Grove Press, 1991), 64.

277 **'spider-webs of intricate relationships seeking a form':** Calvino, *Invisible Cities*, 76.

INDEX